# Quantum Unlocked: The Strange Science Beneath Everything

C. Louis-Charles, PhD

Cybersoft Publishing LLC

Fort Washington, MD 20744

Drclaude.net

First Edition, May 2026

# Contents

# Introduction

You are holding a case file.

Not a case file about crime or courtroom drama, but something more quietly astonishing: a systematic record of experiments that forced physicists to conclude, against every intuition they possessed, that the universe at its smallest scale does not behave the way anything else does. Particles exist in multiple states at once. Measuring one object instantly shapes what you find in another, regardless of the distance between them. A particle can pass through a barrier that it has no energy to cross. And the act of looking changes what is being observed.

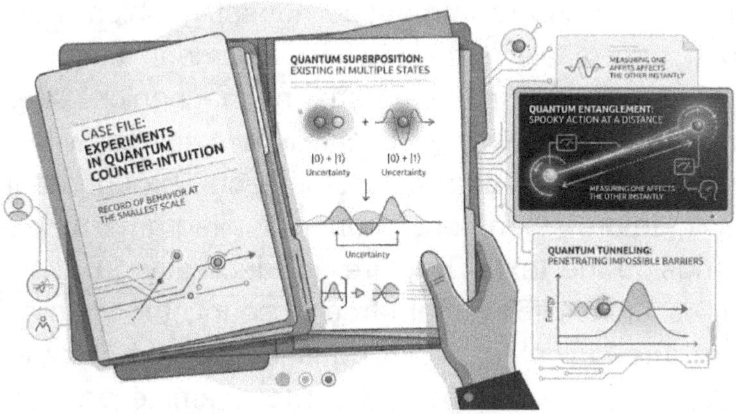


These are not metaphors. They are the literal, experimentally verified behavior of matter and light at the quantum scale. Every result described in this book has been confirmed by physicists working in laboratories on every continent. The evidence is not in dispute. What remains in dispute, even now, is what the evidence means. That is the investigation you are joining.

## Who This Book Is For

This book was written for curious adults with no background in physics who have reached the point where they refuse to nod politely at words like quantum supremacy or wave function without knowing what they mean. Whatever brought you here, one thing is not required: mathematical training. There is no math in this book. Not a single equation appears in its pages. This book explains every concept that physicists convey through equations, words, diagrams, and carefully chosen analogies.

The analogies are introduced honestly, which means each one is also corrected the moment it starts to mislead. Because in quantum mechanics, every analogy eventually fails. That failure is not a flaw in the analogy. It is a feature of the subject. Throughout the book, more than one hundred diagrams appear alongside the text. They are not decorative. Because this is an explicitly math-free account, the diagrams carry the logical load that equations would carry in a textbook. You are encouraged to examine each one carefully before reading the text that follows.

## The Role You Play

From the first chapter to the last, you are positioned as an investigator. Not a student being handed conclusions to memorize, but a careful reasoner confronting a body of experimental evidence and asked to follow it wherever it leads. The chapters open with physical scenarios and historical moments presented as mysteries. Something does not add up. The evidence points in an uncomfortable direction. Your job is to notice that, and then to reason toward the most defensible conclusion the evidence will support.

This framing reflects how quantum physics was actually discovered. The physicists who built quantum mechanics in the early twentieth century were not theorists who invented a new picture of reality from speculation. They were investigators who had run out of classical explanations. The evidence had backed them into a corner, and quantum mechanics was the only exit. The theory was not chosen because it was elegant. It was chosen because it worked, and because nothing else did. No prior knowledge of physics is assumed at any point. Every term is defined the first time it appears, and every concept is built from the ground up before the next one is introduced.

## The Structure of the Investigation

The book proceeds through ten chapters, each focused on a distinct piece of the quantum puzzle, and is best read in order. Chapter One opens where quantum physics actually began: with the failure of classical physics. For two centuries, Newtonian mechanics was the most successful intellectual achievement in human history. Then, at the turn of the twentieth century, physicists pointed their instruments at atoms, and the rules stopped working. Understanding why classical physics failed is the only honest starting point for understanding what replaced it.

Chapters Two through Six build the core vocabulary, one concept at a time. Quantization, the strange discreteness of energy at the atomic scale. Wave-particle duality through the double-slit experiment, which Richard Feynman called the central mystery of quantum mechanics. The uncertainty principle is a feature of reality itself, not a limitation of instruments. Superposition as a genuine physical state, not a metaphor for ignorance. And the wave function, stripped of mathematical notation down to its plain meaning: a complete accounting of a particle's possible futures.

Chapters Seven and Eight extend the investigation into territory that troubled even the physicists who discovered it. Quantum entanglement, which Einstein called spooky action at a distance, is real and experimentally confirmed. Quantum tunneling, the ability of particles to pass through barriers they lack sufficient classical energy to overcome, is the mechanism that powers the sun and drives the radioactive decay of atoms. Chapter Nine is the payoff: every major technology of the digital age, from transistors to lasers to MRI scanners to the quantum computers now emerging from research laboratories, is a quantum machine.

Chapter Ten is an honest account of what quantum mechanics still cannot tell us. The Copenhagen Interpretation, the Many Worlds Interpretation, and pilot wave theories all agree on every experimental prediction and disagree, fundamentally, on the nature of reality. That disagreement is not a failure of science. It is an invitation to take your own position on the frontier.

## A Note on Difficulty

Quantum mechanics is genuinely strange. That strangeness is not a failure of explanation. It is a property of the subject. The universe at the quantum scale does not behave like anything at the human scale, and no amount of careful analogy will make it feel entirely familiar. What careful analogy can do is make it precise. By the end of this book, you will not feel that quantum mechanics is merely mysterious. You will feel that it is specifically strange, in ways you can name, describe, and reason about. That is the difference between mystification and comprehension, and it is the only outcome this book is designed to produce. The investigation begins now.

# 1 The World That Broke: Why Classical Physics Ran Out of Answers

Picture yourself as an investigator arriving at a crime scene that everyone insists is routine. The floor plan is familiar. The evidence is cataloged. Your colleagues have solved a thousand cases just like this one, and the tools in your bag have never let them down. Then you crouch beside the evidence, peer through your lens, and realize something is deeply, irreversibly wrong. The rules that governed every previous case do not apply here. The evidence refuses to fit any known pattern. The case is not routine at all.

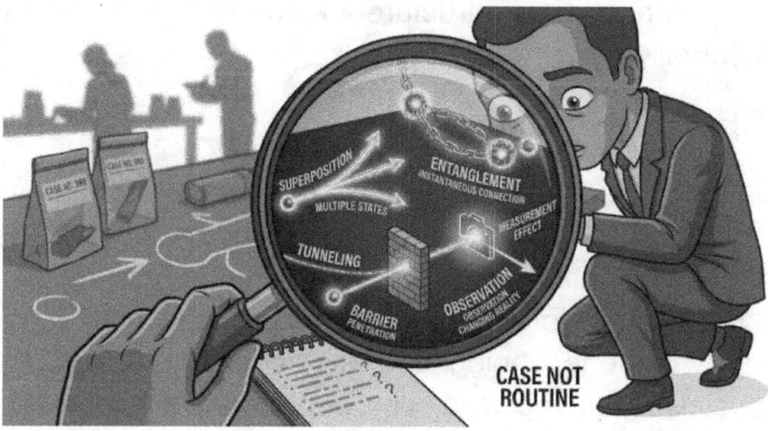


That is precisely the position physicists found themselves in at the turn of the twentieth century. For more than two hundred years, they had carried a set of intellectual tools that worked everywhere they pointed them. Planets, pendulums, bridges, steam engines, ocean waves, cannon trajectories: the framework built by Isaac Newton and refined by a parade of brilliant successors answered every question with stunning precision. Then, researchers turned their instruments toward atoms, hot glowing objects, and the ejection of electrons from metal surfaces, and the tools stopped working. Not because the tools were carelessly applied. Not because the experiments were flawed. The tools simply and definitively broke.

This chapter is the story of that breakdown. It is also the story of the very first, almost accidental steps toward a replacement. Understanding why classical physics failed is not a detour from the real story. It is the real story. The anomalies that stumped the best minds of 1895 to 1910 are the cracks through which the entire strange structure of quantum mechanics eventually grew.

## Why This Matters

Quantum mechanics is not an obscure corner of theoretical physics. It is the foundation beneath semiconductors, lasers, MRI machines, solar panels, and the transistors inside every device you own. Every one of those technologies exists because a handful of physicists refused to sweep experimental anomalies under the rug. They took the broken results seriously, followed the evidence wherever it led, and eventually arrived at a description of nature so strange that even its inventors found it unsettling.

Before we can understand what quantum mechanics says, we need to understand what it replaced and why the replacement was necessary. Classical physics was not a collection of lucky guesses. It was a magnificent intellectual achievement that correctly described an enormous range of phenomena. The fact that it broke down tells us something profound: the universe has layers, and the rules that govern everyday objects are not the same as those that govern atoms. Recognizing where the old rules worked and where they shattered is the first step toward genuine understanding.

## The Newtonian Triumph: A Framework That Explained Everything

To appreciate how catastrophic the breakdown was, you first need to feel the full weight of what Newton's framework achieved. In 1687, Newton published his Principia Mathematica, and with it came something that had never existed before: a single, compact set of rules that could predict the behavior of any moving object in the universe, from apples to planets.

Newton's first law states that an object at rest stays at rest, and an object in motion keeps moving unless acted upon by an outside force. His second rule said that force equals mass times acceleration: push harder on a heavier object, and it accelerates less. Every engineering calculation you have ever relied on, from the design of bridges to the trajectory of satellites, traces back to this close relationship. His third rule, that every action has an equal and opposite reaction, explains how cannons recoil, and rockets rise. Together with Newton's law of universal gravitation, these rules gave physicists a machine for predicting the future of any object if they knew its current position, speed, and the forces acting on it.

Diagram 1.1 - Timeline of Classical Physics Triumphs vs. Experimental Anomalies

Classical physics triumphed for two centuries, then stalled at the atomic scale.

Newton's framework expanded dramatically in the centuries that followed. In the 1800s, James Clerk Maxwell unified electricity, magnetism, and light into four elegant equations. His framework predicted that light was an electromagnetic wave, and experiments confirmed this with extraordinary precision. Thermodynamics grew into a rigorous discipline, giving engineers the tools to calculate the efficiency of steam engines and to trace energy from one form to another. The tools were reliable enough to build the Industrial Revolution on.

By the 1880s, there was a genuine sense in the physics community that the great work was nearly done. The major laws were in place. What remained, some believed, was to measure physical constants more precisely and to apply known laws to unexplored situations. One prominent physicist reportedly told his students that physics was essentially finished: only two small clouds remained on the horizon. Those clouds turned out to be thunderstorms.

## The Assumptions Beneath the Triumph

Every powerful framework rests on assumptions. The assumptions of classical physics were so deeply embedded, so constantly confirmed by everyday experience, that most physicists never thought to question them. Examining those assumptions now is like an investigator going back to review evidence everyone else dismissed as irrelevant.

The first assumption was continuity. In classical physics, energy, motion, and all physical quantities change in smooth, continuous ways. If you push a ball a little harder, it moves a little faster. If you heat a gas gradually, its temperature rises gradually. There are no jumps, no gaps, no forbidden intermediate values. You can always find a quantity between any two values you have already measured.

The second assumption was determinism. Given perfect knowledge of a system's current state, classical physics could in principle predict its entire future. The universe, from this perspective, was a mechanism, vast and intricate but fundamentally clockwork. Every future state followed necessarily from every past state.

There was no room for genuine randomness, only for ignorance.

The third assumption was locality. Objects interact through fields and forces that propagate at finite speeds through space. An event in one place cannot instantly affect a distant place. This assumption was so natural that it barely seemed like an assumption at all. It matched every experiment and every experience.

The fourth assumption was separability. The properties of a system, its position, speed, energy, and temperature, were distinct, independently measurable quantities. You could pin down where something was and how fast it was moving at the same time, in principle at least, even if your instruments were imperfect. The properties were real, definite, and simultaneous.

Diagram 1.2 - The Four Pillars of Classical Physics

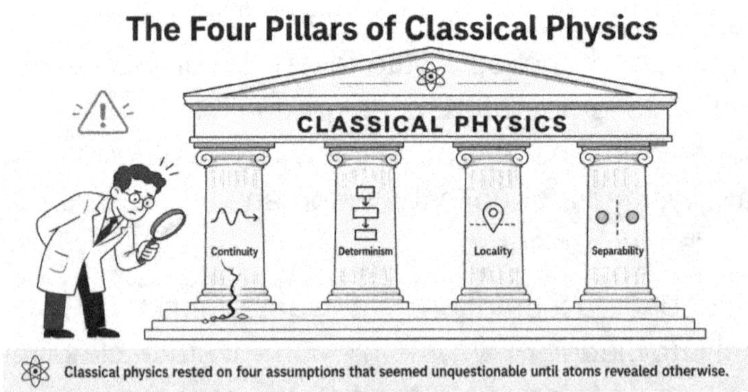


These four assumptions worked perfectly at the scales humans inhabit: planets, pendulums, gas tanks, water pipes. At those scales, they are not merely approximately correct. They are correct to a degree of precision that exceeds any instrument we can build. The crisis that arose at the end of the nineteenth century did not reveal a hidden flaw in classical physics at everyday scales. It revealed that the framework, no matter how well-tuned, did not apply at the atomic scale. The territory had changed, and the map was useless.

## The Map Analogy

Imagine you are a cartographer who has spent decades making extraordinarily accurate maps of Europe. Every coastline, every river bend, every mountain pass is precisely recorded. Then someone hands you an assignment to chart the interior of a single grain of sand. You unroll your tools and begin, only to discover that the usual rules of geometry dissolve at that scale. Boundaries that were crisp and definite at large scales turn out to be fuzzy and probabilistic at small ones.

This analogy captures something real, but it is worth noting where it fails: a grain of sand is still large enough for classical physics to describe it reasonably well. The true breakdown happens at the scale of individual atoms and electrons, roughly a hundred million times smaller than a millimeter. At those scales, the universe behaves in ways that no map drawn with classical tools can represent.

Diagram 1.3 - The Map Analogy: Classical Tools in Quantum Territory

## Blackbody Radiation: The First Crack

Let us now examine the specific experimental results that challenged the classical framework. The first involves something everyone has experienced: an object that glows because it is hot.

Heat a piece of metal, and it first glows dull red, then bright orange, then yellow, then white. This progression of color with temperature is so familiar from campfires, stovetops, and sunsets that it seems almost too ordinary to investigate. But physicists in the late 1800s asked a precise version of this question: if you had a perfectly efficient heat radiator, one that absorbed every wavelength of light and emitted radiation purely based on its temperature, what would its spectrum look like?

This idealized object is called a blackbody. The word is counterintuitive because a glowing hot object certainly does not look black. The name refers to its absorbing properties: a perfect blackbody absorbs all incoming radiation without reflecting any. At room temperature, such an object would appear perfectly black because it would reflect nothing. But heat it, and it radiates with a characteristic spectrum that depends only on temperature, not on what the object is made of.

Experimenters in the 1890s carefully measured this spectrum and found a distinctive shape. At low frequencies, the emitted power rises with frequency. It reaches a peak that depends on temperature, then falls back down. As you heat the object, the peak shifts toward higher frequencies, which is why the color shifts from red toward white as the temperature increases. The total radiated power also grows rapidly with temperature.

Diagram 1.4 - Blackbody Radiation Curve: Classical Prediction vs. Observed Reality

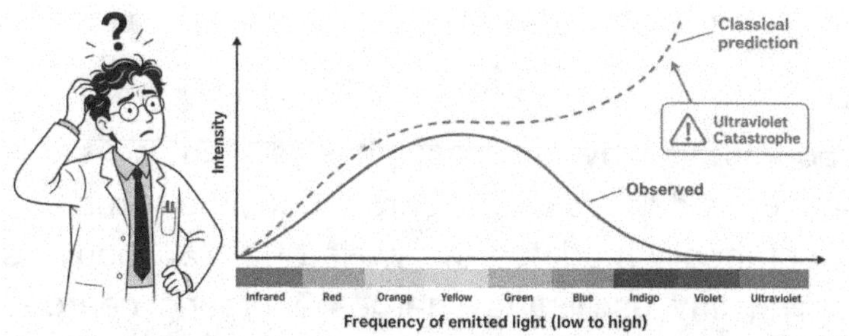


**Classical theory predicted infinite brightness. Nature politely but firmly disagreed.**

Physicists applied classical physics to predict the shape of the blackbody spectrum. They had powerful tools: Maxwell's electromagnetic theory, thermodynamics, and the assumption that energy was continuous. When they cranked through the calculation, they got a result for the low-frequency end of the spectrum that matched experiments beautifully. Encouraged, they extended the calculation to higher frequencies, into the ultraviolet region and beyond.

The prediction was catastrophic. Classical theory said the intensity of radiation should continue to rise without limit as the frequency increased. A glowing object should be infinitely bright at ultraviolet and higher frequencies. This prediction became known as the ultraviolet catastrophe, and it represented a complete failure of the classical framework. It was not a small discrepancy. It was an infinite one. Every hot object in the universe should be emitting infinite energy according to classical physics, and yet the universe around you is distinctly finite.

## Why Classical Theory Predicted a Catastrophe

To understand why classical physics went so badly wrong, consider how it approached the problem. Classical theory modeled the walls of a blackbody as containing tiny oscillating charges: vibrating charged particles that could absorb and emit electromagnetic radiation. The theory assumed that any oscillator could take on any energy value whatsoever, from zero upward, in a continuous, smooth range.

When physicists counted how many different oscillation modes could fit in a cavity and multiplied that by the available energy, they found that higher-frequency modes vastly outnumbered lower-frequency modes. Because classical theory assigned equal average energy to every mode, the prediction followed mechanically: more modes at high frequencies meant more radiated energy at high frequencies, rising without bound. The math was correct given the assumptions. The assumptions were wrong.

Diagram 1.5 - Why the Ultraviolet Catastrophe Arises: Counting Oscillation Modes

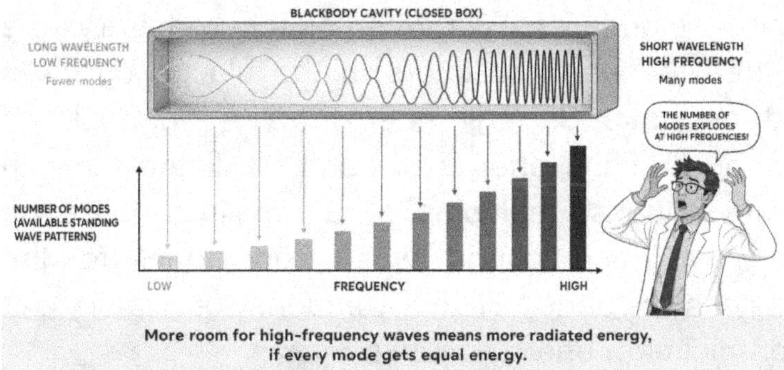


This was the situation facing Max Planck in 1900. He was not a young rebel looking to overturn the established order. He was a careful, conservative physicist who believed deeply in classical thermodynamics and was frustrated by this particular anomaly. He set about deriving a formula that would match the observed blackbody spectrum, not because he wanted to challenge classical physics, but because he thought he could rescue it.

## Max Planck's Desperate Mathematical Fix

Max Planck spent months wrestling with the blackbody problem. He tried various approaches within the classical framework. None of them produced a formula that matched the data across all frequencies. Then he tried something he himself described as a desperate act.

Planck made a mathematical assumption that, by classical standards, was completely arbitrary: he assumed that the oscillators in the blackbody walls could not take on any energy value. Instead, he assumed that the energy of each oscillator came in discrete chunks, multiples of a smallest unit. The energy of each chunk was proportional to the oscillator's frequency. Higher-frequency oscillators had larger minimum energy chunks.

The formula this produced matched the experimental data perfectly across the entire spectrum. Every peak, every slope, every detail fell exactly into place. Planck had his formula. He was deeply uncomfortable with the assumption that gave rise to it.

Here is the physical meaning of what Planck had done, stripped of the mathematics. High-frequency oscillators have large minimum energy chunks. Because thermal energy is limited, only a few of those high-frequency oscillators can afford to vibrate at all. Most of them sit silently because they lack the energy to pay the minimum admission price, as it were. This naturally cuts off the radiation at high frequencies and eliminates the ultraviolet catastrophe. The formula works because it introduces a fundamental discreteness into energy exchange.

Planck's constant, the tiny number that sets the size of these energy chunks, was born in that calculation. Its value is extraordinarily small, about 6.6 × 10^-34 in standard units. This is why quantum effects are invisible in everyday life: the energy chunks are so small compared to the energies of ordinary objects that they appear to form a continuous spectrum. It is only at the atomic scale, where the energies involved are comparable in magnitude to those of these tiny chunks, that the discreteness becomes detectable.

Diagram 1.6 - Planck's Quantum Fix: Energy Chunks Instead of a Continuous Ladder

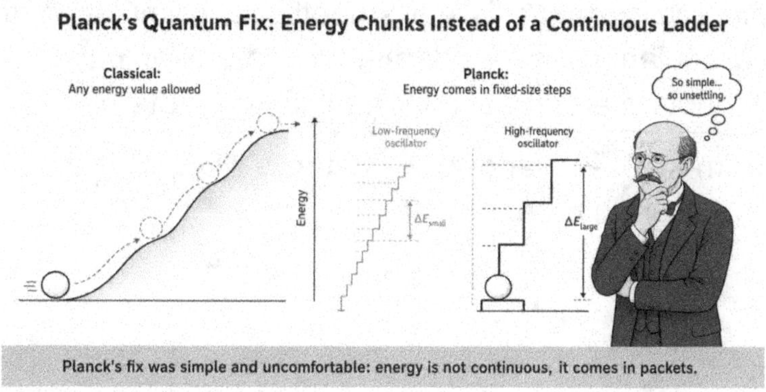

Planck himself hoped that the discreteness was a mathematical trick with no physical reality. He hoped that a more rigorous classical derivation would eventually reproduce his formula without requiring discrete energy chunks. That hope was extinguished five years later, when Albert Einstein applied Planck's idea to a completely different problem and found that the energy chunks were not just mathematically convenient. They were physically real.

## What Planck's Constant Tells Us

Planck's constant is one of the most important numbers in physics. It appears in virtually every quantum calculation, setting the scale at which classical physics breaks down. Its smallness explains why quantum effects are invisible in everyday life; its finiteness explains why atoms are stable rather than radiating away all their energy instantaneously. Think of it as setting the universe's minimum grain size: below a certain scale, nature is not smooth but granular, and that granularity is a fundamental feature of physical reality, not an artifact of imperfect instruments.

Diagram 1.7 - Planck's Constant: The Universe's Minimum Grain Size

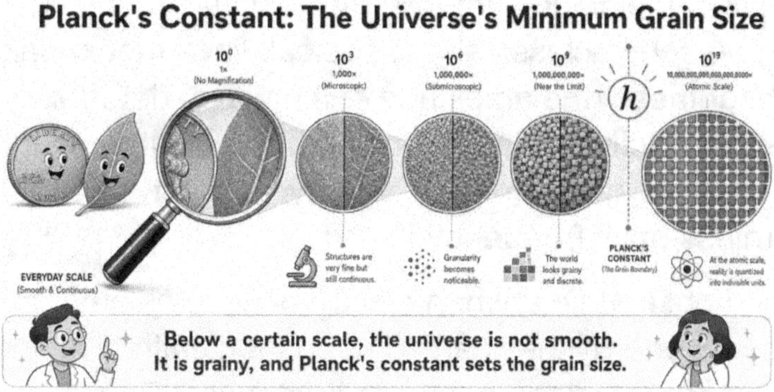


## Photoelectric Effect: Einstein Confirms the Chunks Are Real

The photoelectric effect had puzzled experimenters since 1887, when Heinrich Hertz observed that ultraviolet light could cause metal surfaces to emit sparks. Subsequent work showed that light shining on a metal surface could eject electrons from it. This became the standard experimental setup: shine light on a metal, measure whether electrons are emitted, and carefully vary the properties of the light to see what changes.

The results were deeply strange from a classical standpoint. Here is what investigators found, laid out as evidence at a crime scene.

First observation: whether electrons were emitted at all depended on the light's frequency, not on its brightness. Red light, no matter how intense, ejected no electrons from certain metals. Ultraviolet light, even faint ultraviolet light, immediately ejects electrons. This made no sense classically. Classical wave theory said that brightness (the amplitude of the wave) determined the energy delivered to the metal, and a bright enough wave should have enough energy to eject electrons regardless of its frequency.

Second observation: when electrons were ejected, the maximum energy they carried depended entirely on the frequency of the light, not on its brightness. Double the brightness, and you get twice as many electrons ejected, but each electron carries the same energy as before. Increase the light frequency, and each ejected electron carries more energy. Decrease the frequency and each carries less.

Third observation: there was a sharp threshold frequency below which no electrons were ejected at all, no matter how bright the light. Below this threshold, nothing. Above it, ejection happened instantaneously, even in the dimmest light.

These observations were completely incompatible with classical wave theory. According to classical physics, light was a smooth, continuous wave. A brighter wave carried more total energy. Given enough time, even a low-frequency wave should be able to shake electrons loose from a metal surface. There was no reason in classical theory for a sharp frequency threshold, and no mechanism by which frequency could determine the energy of individual electrons.

Diagram 1.8 - The Photoelectric Effect: What Classical Theory Expected vs. What Actually Happened

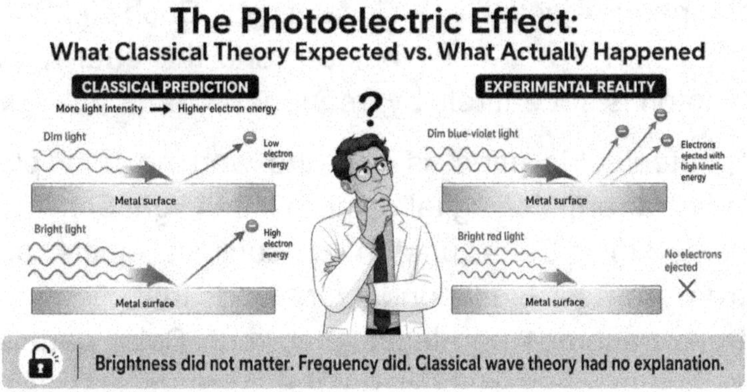


## Einstein's Radical Solution

In 1905, the same year he published his theory of special relativity, Albert Einstein published a paper on the photoelectric effect. His solution was both simple and radical: light is not a continuous wave. Light comes in discrete packets of energy, just as Planck's formula implied. Each packet carries an energy determined by its frequency. Einstein called these packets light quanta. We now call them photons.

Einstein's explanation flows directly from this idea. When a photon strikes a metal surface, it delivers its entire energy to a single electron in one indivisible interaction. If the photon's energy falls below the binding energy of the electron, nothing happens. If it exceeds the threshold, the electron breaks free and carries the leftover energy as kinetic energy. This explains every feature of the data: the sharp frequency threshold, the dependence of electron energy on frequency rather than brightness, and the instantaneous ejection even in faint light. Each electron interacts with exactly one photon, and the energy of that photon is set entirely by frequency.

The photoelectric effect is now the working principle behind solar panels, digital cameras, and light sensors of every kind. Each photon of sunlight ejects an electron from a semiconductor, and that flow of electrons becomes an electrical current. Solar panels are quantum devices built on the ruins of classical wave theory.

Diagram 1.9 - Einstein's Photon Model: Light as Discrete Packets Striking Metal

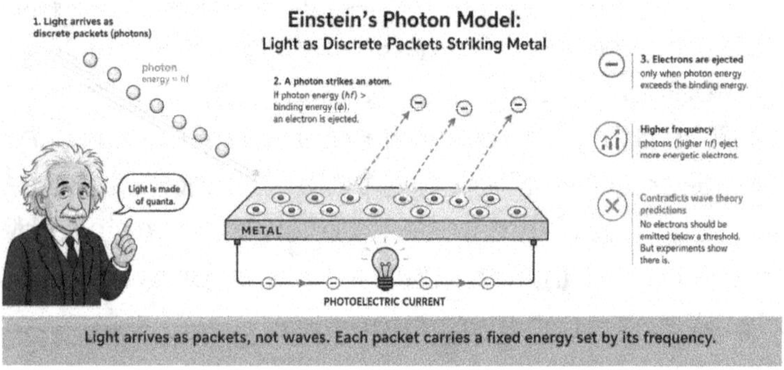


Einstein won the Nobel Prize in 1921 specifically for the photoelectric effect, not for relativity, which tells you something about how significant this result was to the physics community. Relativity was revolutionary in its implications for space, time, and gravity. But the photoelectric effect was revolutionary in a different, arguably more fundamental way: it meant that light, which Maxwell had definitively shown to be a continuous electromagnetic wave, was also, in some sense, a stream of discrete particles. This contradiction would take decades to understand fully, and the tools developed to address it would become the foundations of quantum mechanics.

### Solar Panels: Classical Physics Could Not Have Designed Them

Solar panels are one of the most consequential technologies of the twenty-first century, and they are quantum at their heart. A classical physicist who understood only wave theory would have no framework for designing a photovoltaic cell. Continuous waves do not have a threshold frequency. The design principle that makes a solar panel work, selecting a semiconductor with the right energy gap to match the frequency of incoming photons, is entirely quantum mechanical. The photoelectric effect was discovered by Hertz in 1887, explained by Einstein in 1905, and transformed into a gigawatt-scale energy technology over the following century. The quantum revolution did not begin in abstract theory. It began with measurements that did not fit, and it ends with devices that power cities.

## The Limits of Continuous Models

Let us step back from the specific experiments and look at the general pattern. Both blackbody radiation and the photoelectric effect exposed the same fundamental weakness in classical physics: the assumption that energy is continuous.

Classical physics assumes you can always find a quantity between any two values. There is no smallest unit of energy, no minimum step size, no forbidden gap between one level and the next. Everything flows smoothly. This assumption is so deeply embedded in classical mathematics that it forms the foundation of calculus itself. Calculus works precisely because physical quantities can be subdivided indefinitely.

At everyday scales, this assumption works spectacularly well. The energy difference between adjacent quantum states for a macroscopic object is so fantastically small that no instrument could detect it. The object behaves as if energy were continuous because the steps between quantum states are invisible. Classical physics is a precise and accurate limiting case of the deeper quantum theory, valid whenever objects are large enough that their quantum graininess becomes imperceptible.

But at the atomic scale, the steps between energy levels are comparable to the energies involved in atomic processes. When an electron absorbs or emits energy, it can do so only in amounts that match specific transitions between specific levels. It must jump or remain still; there is no gradual climb. This has no classical analog and is the source of some of the most distinctive phenomena in the quantum world.

Diagram 1.10 - Continuous vs. Discrete Energy: Everyday Objects vs. Atoms

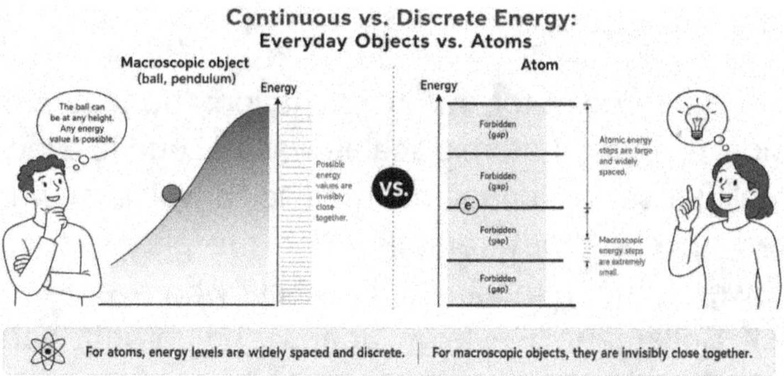


The failure of continuous models at atomic scales is not a limitation of our measuring instruments. It is a fact about nature. Atoms are genuinely discrete systems. Electrons bound to an atom cannot hover at arbitrary distances from the nucleus the way a satellite hovers at arbitrary altitudes above Earth. They occupy specific, fixed shells, and the energy differences between those shells correspond to specific energy packets.

This discreteness has a profound consequence: light emitted by atoms comes in specific discrete frequencies, not a continuous spectrum. Each transition between energy levels produces a photon of a specific frequency. This is why neon signs glow in particular colors, why fireworks produce characteristic bursts of color, and why astronomers can identify which elements are present in a distant star by analyzing the frequencies of its light. All of it traces back to the discreteness Planck first stumbled upon while trying to save classical physics.

## Why the Breakdown Was Unavoidable

Here is a key point for our investigation: the failure of classical physics at atomic scales was not due to a lack of intelligence or imagination. The physicists who worked on blackbody radiation and the photoelectric effect were among the finest scientists who ever lived. Planck, Boltzmann, Rayleigh, Jeans, and Lorentz were not mediocre thinkers who missed obvious answers. They applied the best tools available with extraordinary skill and found that those tools did not apply.

This tells us something important about scientific progress. The failure was not a bug. It was a signal. It told physicists that they had reached the edge of their current map and that a new map was needed. The anomalies were not embarrassing failures to be quietly set aside. They were clues pointing toward a deeper level of description. Every great scientific revolution begins with exactly this pattern: reliable tools that stop working, accumulating anomalies, and a growing recognition that the framework itself needs to be replaced.

The classical framework was replaced not because someone clever thought of a better alternative from scratch, but because experimental evidence demanded it. Quantum mechanics is, in this sense, the most empirically driven theory in the history of science. Every one of its strange features exists because the experiment forced it into existence. It is not a product of theoretical speculation. It is a product of evidence.

## Connecting the Evidence: Classical Physics Was Not Wrong, It Was Incomplete

By 1910, the picture was clear enough to state plainly. Classical physics was not wrong. It was incomplete. It correctly described matter and energy at everyday scales, where quantum effects average out to the point of invisibility. It failed at atomic scales, where the discreteness of energy and the quantum nature of light became decisive.

This distinction matters. Classical physics is not the enemy of quantum physics. It is a special case of it, valid in a certain regime. When physicists say classical physics breaks down at atomic scales, they do not mean Newton's laws are false for a falling apple. They mean those laws are a reliable approximation that becomes exact for large, slow-moving objects but fails for electrons, photons, and atoms.

A flat map drawn on paper works perfectly for navigating within a city or even across a country. The curvature of the Earth is too small to matter at those scales. But plot a course from New York to Tokyo on a flat map, and you will end up in the wrong ocean. The map is not wrong for short journeys. It simply fails at larger scales, where Earth's curvature becomes decisive. Classical physics is a similar approximation, valid at large scales and failing at small ones. The underlying reality is quantum-mechanical, and classical physics emerges from quantum mechanics in the regime where Planck's constant is negligible.

Diagram 1.11 - The Hierarchy of Physical Theories: Where Classical Physics Lives Within Quantum Mechanics

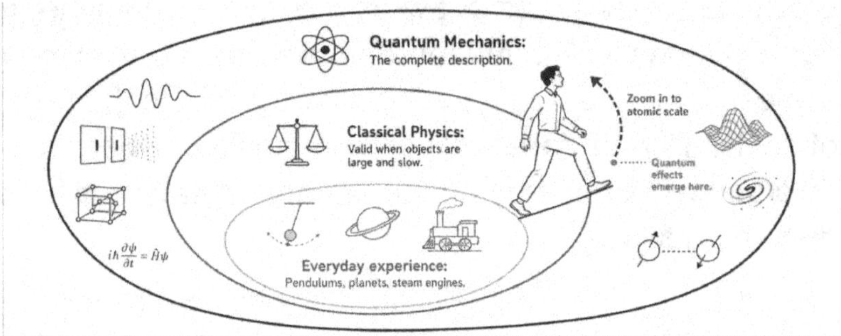


Classical physics is not wrong. It is the quantum world seen from very far away.

You are not being asked to abandon your classical intuitions. You are being asked to recognize their domain of validity and to understand why, when we enter the atomic domain, new and stranger rules are needed. Those rules were not invented. They were discovered, forced out of hiding by experimental evidence that refused to fit any other explanation.

The investigators, spread across many laboratories in many countries, did not start by believing classical physics was wrong. They started by trying to explain anomalies within the classical framework. They failed, methodically and honestly. That failure is one of the most important episodes in the history of science. The breakdown of classical physics is not a story about scientists being wrong. It is a story about scientists who are honest enough to follow the evidence wherever it leads, even when it leads to something profoundly uncomfortable.

## Takeaway

Over time, theorists did not invent quantum physics to speculate. It was forced upon physicists by experimental evidence that refused to fit any other explanation. Two precise experiments, blackbody radiation and the photoelectric effect, each shattered a foundational assumption of classical physics: that energy is continuous and infinitely divisible.

Max Planck's discovery that energy comes in discrete chunks was an act of mathematical desperation that revealed a deep truth about nature. Albert Einstein's explanation of the photoelectric effect, that light itself consists of indivisible packets, confirmed that discreteness was real, not a computational trick. Together, these discoveries marked the birth of quantum mechanics, not as a theory but as a recognition that the universe at atomic scales operates by different rules than the universe we can see and touch.

Classical physics was not wrong. It was incomplete. It is a precise, reliable description of nature at scales where quantum graininess averages to invisibility. The quantum revolution did not discard classical physics. It placed it inside a larger, stranger, more accurate description of reality.

As you move through the rest of this book, you are the investigator who has just examined the first piece of anomalous evidence. You know that the familiar map has limits. You know that a new kind of map is needed. In the chapters ahead, you will help build it.

# 2 The Packet Problem: How Energy Stopped Being a River and Became a Rain

Picture yourself standing at a kitchen sink, letting the tap run. The water pours out in a smooth, unbroken stream. You can turn the valve to any position and get any flow rate you like, from a thin trickle to a full blast. The stream is continuous. There is no smallest unit of flow you are forced to accept. The water does not arrive in little lumps with gaps between them.

For two centuries, physicists assumed that energy behaved exactly like water. Light radiating from a hot coal, heat conducted through a metal rod, the vibrations of atoms in a crystal, all of it was imagined as a smooth, continuous river that could carry any amount of energy you cared to specify. The laws of classical physics were built on that assumption. They worked brilliantly for everything from cannon trajectories to electrical circuits. Then, around the turn of the twentieth century, a handful of stubborn experimental results refused to fit the picture. The river, it turned out, was not a river at all. It was raining, made of individual drops. Those drops were quanta, and their discovery changed the foundation of all physics.

In this chapter, you are the investigator who follows the trail of those anomalies. You will meet the blackbody problem that launched the quantum idea, the photoelectric effect that cemented it, and the strange spectral fingerprints of atoms that made quantization not just convenient but mandatory. By the end, you will be able to look at a neon sign or an LED light and see, in the colored glow, the signature of atoms trading energy in fixed, indivisible packets.

## Why This Matters

The concept of quantization is not a minor technical detail. It is the bedrock on which all of modern physics and chemistry rest. Without it, atoms would be unstable, electrons would spiral into nuclei in a fraction of a second, and matter as we know it would not exist. Every LED display, every laser pointer, every MRI machine, and every silicon chip in every computer operates because energy at the atomic scale comes in discrete steps rather than a continuous flow. Understanding what a quantum actually is, not as vague jargon but as a physical reality grounded in experiment, is the essential first step into the quantum world.

## The Classical Picture: Energy as a River

To appreciate why quantization was so shocking, you first need to understand what physicists expected. The classical picture of energy is deeply intuitive because it matches everyday experience almost perfectly.

### Continuous Quantities in Everyday Life

Think about adjusting the brightness of an old-style incandescent lamp with a dimmer switch. You turn the dial slowly from off to full brightness, and the light increases smoothly, without jumps. You can set it to any level you want. Classical physics described heat and light in exactly this way. Energy was a continuous quantity, like the position of that dimmer switch, able to take any value on a smooth, unbroken scale.

The mathematical framework that codified this picture was James Clerk Maxwell's theory of electromagnetic radiation, completed in 1865. Maxwell showed that light was a wave, an oscillating combination of electric and magnetic fields propagating through space. Like all waves, it had a frequency, a wavelength, and an amplitude. In classical theory, the energy carried by a wave depended on its amplitude: a bigger wave carried more energy, and you could make the amplitude as large or as small as you pleased. There was no minimum packet size. Energy, like water in a river, could be divided arbitrarily finely.

Diagram 2.1 – The Classical Continuous Energy Model vs. the Quantum Rain

This classical model predicted many things correctly. Bridges, steam engines, radio transmitters, and optical telescopes all operate in regimes where the continuous-energy assumption holds. The trouble only appeared at the atomic scale, where the amounts of energy involved were small enough that the discrete nature of reality became impossible to ignore.

## The Trouble with Hot Objects

One of the cleanest ways to study electromagnetic radiation is to examine the emission from a heated object. Every physical object at a temperature above absolute zero radiates energy as electromagnetic waves. A metal poker heated in a fire first glows dull red, then bright orange, then approaches white as the temperature rises. The color indicates how energy is distributed across different frequencies.

Physicists invented an idealized version of this setup called a blackbody: an object that absorbs all incident radiation and re-emits it as a function of its temperature alone. A small hole in a hollow metal cavity is a good practical approximation. The radiation leaking through the hole is called blackbody radiation, and measuring its spectrum, how much energy comes out at each frequency, was a straightforward laboratory exercise by the 1890s.

Classical physics made a firm prediction about that spectrum. Using Maxwell's wave theory and the standard rules of statistical mechanics, Lord Rayleigh and James Jeans derived a formula now called the

Rayleigh-Jeans law. It agreed beautifully with the experiment at low frequencies. But at high frequencies, it made an absurd prediction: as you moved toward shorter wavelengths, into violet and ultraviolet light, the predicted energy should keep rising without limit. A glowing object, according to the classical formula, should blast out infinite energy at short wavelengths and immediately destroy everything nearby. Since nothing of the sort happens when you stand next to a campfire, something was clearly wrong. Physicists called this failure the ultraviolet catastrophe.

Diagram 2.2 – The Ultraviolet Catastrophe: Classical Prediction vs. Reality

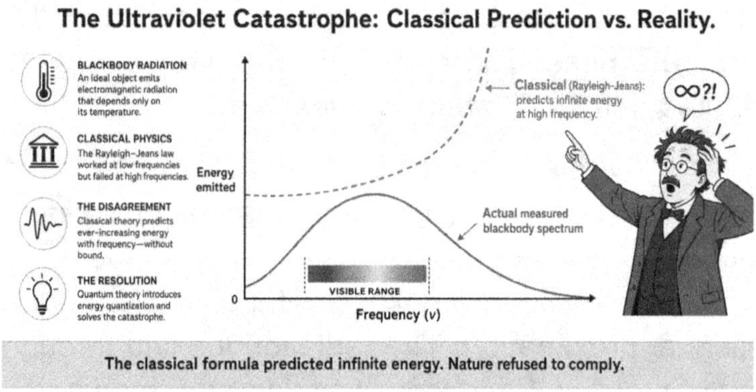


The ultraviolet catastrophe was not a minor discrepancy that could be patched with a small correction. It was a fundamental breakdown of the classical picture. Fixing it would require throwing out one of the core assumptions: the idea that energy is continuous.

## Planck's Desperate Gamble: Inventing the Quantum

In December 1900, the German physicist Max Planck presented a formula to the German Physical Society that fit the blackbody spectrum exactly. It was the right answer. The problem was that Planck had no clear physical reason for the formula. He had essentially worked backward from the data, adjusting his mathematics until the curve matched, and then asked himself: What assumption would make this formula true?

### The Assumption That Changed Everything

The answer Planck discovered was uncomfortable. The only way to make his formula work was to assume that the tiny oscillating atoms in the cavity walls could not absorb or emit energy in any arbitrary amount. Instead, each oscillator could only exchange energy in whole-number multiples of a fixed minimum chunk. The size of that chunk was proportional to the frequency of the oscillator. A high-frequency oscillator had a large minimum chunk; a low-frequency oscillator had a small one.

Planck wrote this mathematically as $E = hf$, where E is the energy of a single packet, f is the frequency, and h is a tiny constant that Planck fitted to the data. That constant is now known as Planck's constant. Its value is approximately 6.626 × 10^-34 joule-seconds. It is an extraordinarily small number, which is why the discrete nature of energy is invisible in everyday life. At human scales, the packets are so tiny that the flow of energy appears perfectly smooth, just as a beach looks smooth even though it is made of individual grains of sand.

Planck himself was not entirely happy with his own solution. He viewed the energy quantization as a mathematical trick, a bookkeeping device rather than a statement about physical reality. He spent years trying to derive the correct blackbody formula without invoking discrete packets, and he never fully succeeded. It would take another physicist, five years later, to insist that the packets were real.

Diagram 2.3 – Planck's Quantum Fix: Discrete Chunks Replace the Continuous River

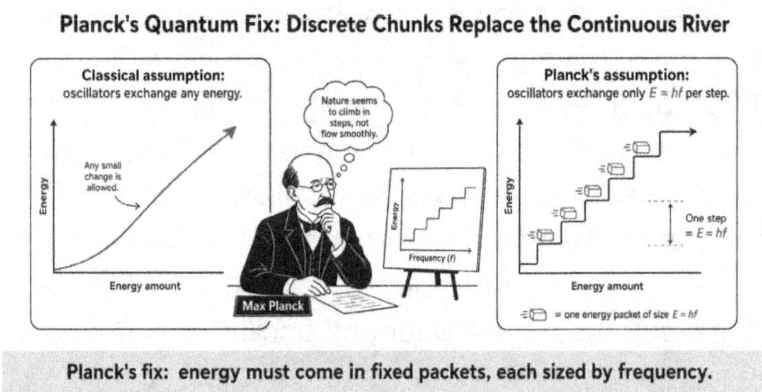


The staircase image is worth pausing on, because it will reappear throughout this book. A staircase forces you to stand at specific heights. You can be on the first step, the second, or the third, but you cannot hover between them. A ramp, by contrast, lets you stand at

any height at all. Classical physics said energy was a ramp. Planck's formula required a staircase. The analogy has its limits, of course: real stairs are made of matter, and you could, in principle, carve a step to any height. Quantum staircases are different. The heights are fixed by nature and cannot be negotiated. That is the deeper point.

## Einstein and the Photon: Light as Packets

Albert Einstein read Planck's work and drew the more radical conclusion that Planck himself had refused to draw. If oscillating atoms could only exchange energy in fixed packets, Einstein argued, then perhaps light itself was made of discrete packets. Light would not be a continuous wave that happened to interact with matter in chunks. The packets would be a real feature of the electromagnetic field.

### The Photoelectric Effect: A Clue That Would Not Go Away

The evidence Einstein used came from the phenomenon known as the photoelectric effect. When light shines on a metal surface, electrons are sometimes knocked loose. This was well known by 1905. What was not well explained was the peculiar pattern of results.

Classical wave theory predicted that the energy of the ejected electrons should depend on the light intensity: brighter light carries more energy and should produce more energetic electrons. It also predicted that with dim enough light, you would have to wait a while for enough wave energy to build up before any electrons popped out, the way a bathtub fills slowly from a dripping tap.

The actual results were completely different. The energy of the ejected electrons did not depend on brightness at all. It depended on the color of the light, i.e., on its frequency. Dim blue light ejects high-energy electrons. Bright red light ejected no electrons at all, or low-energy ones, no matter how intense the red beam was made. And the electrons appeared instantly, without any detectable delay, even under very faint illumination.

Einstein's explanation was simple and devastating. Light consists of individual packets, each carrying an energy equal to h times its frequency. When one packet strikes an electron, it transfers its entire energy in a single transaction, the way a billiard ball transfers momentum when it strikes another. A high-frequency packet has enough energy to eject an electron; a low-frequency packet does not, no matter how many packets per second arrive. Brightness measures the number of packets, not the energy per packet. Einstein called these packets quanta of light. We now call them photons.

Diagram 2.4 – The Photoelectric Effect: Packets vs. Waves

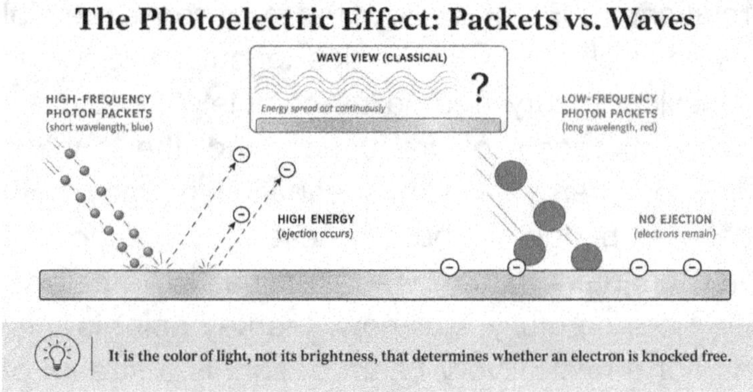


For this explanation, Einstein received the Nobel Prize in Physics in 1921. Note that he did not receive it for relativity, which is the work he is most famous for. The Nobel committee recognized the explanation of the photoelectric effect as foundational to the new quantum physics.

## What a Photon Actually Is

A photon is not a speck of matter. It has no mass. It is a discrete bundle of electromagnetic energy that travels at the speed of light. Its energy is fixed by its frequency: $E = hf$. A photon of red light at a frequency of roughly 430 trillion oscillations per second carries about 1.8 electron-volts of energy. A photon of violet light at roughly 750 trillion oscillations per second carries about 3.1 electron-volts. A photon of a dental X-ray carries thousands of electron-volts. These differences in energy are why UV light causes sunburn and visible light does not, and why X-rays penetrate skin while visible photons bounce off.

The relationship E = hf is one of the most consequential equations in all of physics. It ties together two things that classical theory treated as completely separate: the wave property of frequency and the particle property of energy. A single equation bridges both worlds. This bridge will become even more important in later chapters when we examine wave-particle duality more carefully. For now, the key point is this: light is not a river. Every beam of light is a stream of individual photons, each carrying a fixed amount of energy determined solely by its color.

## Atomic Fingerprints: The Mystery of Emission Spectra

Quantization not only affects how hot cavities emit light. It also controls the light emitted by individual atoms, and the evidence for this was sitting in laboratories for decades before anyone could explain it.

### The Rainbow and Its Gaps

Pass white light through a glass prism, and you get a rainbow: a continuous spectrum stretching from red through orange, yellow, green, blue, and violet. Every visible light frequency is present, and the colors blend smoothly into one another. This is sunlight spread out by frequency, and it is continuous because sunlight comes from a dense, hot plasma where countless atoms are constantly colliding and radiating at all frequencies.

Now try a different experiment. Take a sealed glass tube filled with pure hydrogen gas. Run an electric current through it to energize the hydrogen atoms. Pass the light through the tube that emits through a prism. You do not get a rainbow. You get a series of sharp, bright lines at very specific frequencies, with darkness everywhere else. The pattern is distinctive and reproducible: a red line, a blue-green line, a violet line, and more lines in the ultraviolet. No other gas produces exactly this pattern. Neon produces its own set of lines, sodium its own, calcium its own. Each element's emission spectrum is as unique as a fingerprint.

These spectral lines had been cataloged carefully since the 1850s. Astronomers used them to identify elements in the atmospheres of distant stars thousands of years before anyone had a theory to explain them. The pattern for hydrogen was summarized in a formula by Johann Balmer in 1885, a neat numerical relationship with no physical explanation attached. The question every physicist wanted to answer was: why these specific frequencies and not others? Why does hydrogen emit light at exactly those wavelengths and nowhere else?

Diagram 2.5 – Continuous Spectrum vs. Discrete Emission Spectrum of Hydrogen

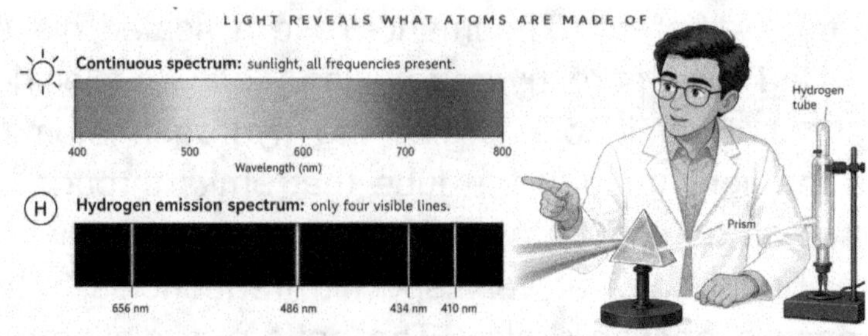


Each element emits a unique set of colors, like a fingerprint written in light.

## Why Neon Signs Glow

This is a good moment to connect the physics to something you see every day. The warm orange-red glow of a neon sign is not a broad, white light filtered through colored glass. It is neon atoms emitting photons at their specific, fixed frequencies. When an electric current runs through the neon gas inside the tube, it excites the neon atoms. As those atoms release their energy, they produce photons at precisely the frequencies determined by neon's unique energy level structure. The collection of those frequencies produces the characteristic reddish-orange color. A sign filled with argon glows blue-violet. Mercury vapor glows blue-white. The colors of the sign tell you exactly which gas is inside the tube, and the reason those colors are so vivid and so pure is that they consist of narrow-frequency photons rather than a broad, smeared-out spectrum.

Fireworks use the same principle. When the shell bursts, the heat excites the metal atoms in the chemical payload. Strontium atoms produce the deep red color because strontium's energy levels are spaced in a way that produces red photons. Copper atoms produce blue-green. Sodium produces a bright yellow. The pyrotechnician choosing chemical compounds is, without knowing quantum physics, choosing emission spectra. The color of a firework is a direct read-out of the atomic energy structure.

Diagram 2.6 – Atomic Color Signatures: Neon Signs and Fireworks

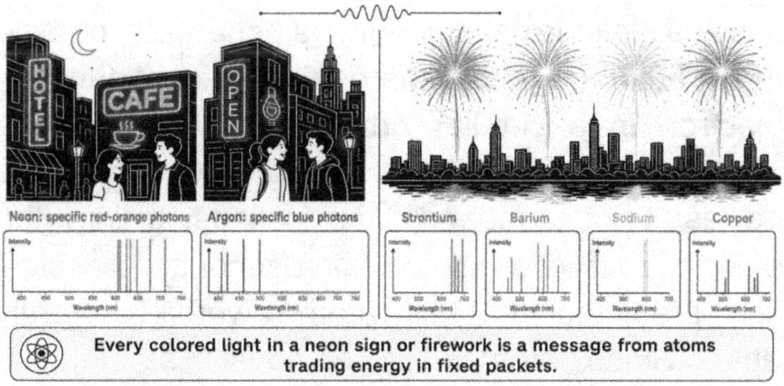


## Bohr's Atom: Electrons on a Staircase

The spectral line mystery was at least partially solved in 1913 by the Danish physicist Niels Bohr. His model of the hydrogen atom was bold and, by the standards of classical physics, completely unjustified. But it predicted the hydrogen spectrum with extraordinary accuracy, and in doing so, it made the quantization of atomic energy levels impossible to deny.

## The Classical Atom and Its Problems

At the time, physicists knew that an atom consisted of a small, dense, positively charged nucleus surrounded by negatively charged electrons. This was the picture established by Ernest Rutherford's gold foil experiments in 1909-1911. The obvious analogy was a miniature solar system: electrons orbiting the nucleus, much like planets orbiting the sun.

The solar system model had a fatal flaw that was immediately obvious to any physicist who looked at it. In classical electromagnetism, a charged particle moving in a curved path radiates energy continuously. An electron in a circular orbit around the nucleus should therefore be constantly losing energy, spiraling inward as it loses speed, and collapsing into the nucleus in a fraction of a microsecond. Classical physics predicted that all atoms were inherently unstable. Since atoms are obviously stable, the classical model was fundamentally wrong. Something was preventing the electrons from spiraling inward. Nobody knew what.

## Bohr's Radical Postulate

Bohr deliberately chose to ignore classical electromagnetic predictions and replace them with a postulate. He declared that electrons in an atom could occupy only specific orbits, and that, while in those orbits, they did not radiate energy, despite what classical physics predicted. Each allowed orbit corresponded to a specific energy level. The electron could sit on level 1, level 2, level 3, and so on upward, but it could not sit between levels. The energy levels formed a staircase, not a ramp.

The electron could jump from one step to another, but only by absorbing or emitting a photon whose energy exactly matched the difference between the two steps. If the electron jumped down from level 3 to level 2, it emitted a photon with energy equal to the gap between those levels. Because the gaps were fixed, the photon had a fixed energy, which meant a fixed frequency and a specific color of light. The collection of all possible downward jumps produced the collection of spectral lines.

Diagram 2.7 – The Bohr Atom: Electrons Jumping Between Energy Levels

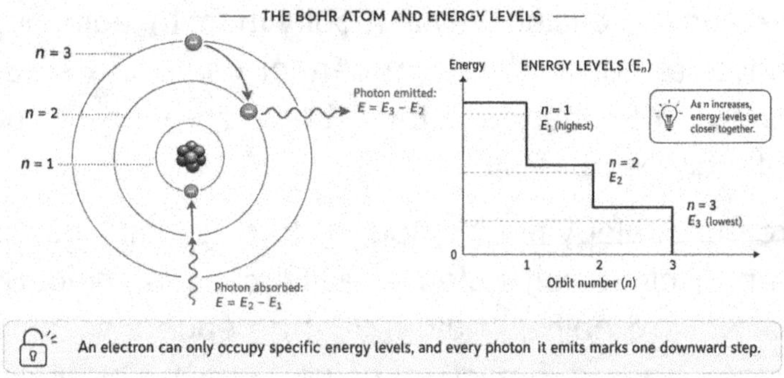


Bohr calculated the energy of each level using a simple formula that incorporated Planck's constant. When he plugged in the numbers for hydrogen, the predicted spectral lines matched the observed ones to within a fraction of a percent. The Balmer formula that had sat

unexplained for nearly thirty years fell out of Bohr's model as a direct mathematical consequence. That agreement between theory and experiment was not a coincidence. It was evidence that the energy levels were real.

## The Staircase vs. the Ramp

This is the moment to look more carefully at the staircase analogy, because it captures the physical reality quite well while having a clear limitation worth naming. A real staircase is made of matter. In principle, you could machine the steps to any height you wanted. You could also slide a thin wedge under one end and make a very nearly ramp staircase. The quantum mechanical staircase is different in kind: the step heights are determined by fundamental constants of nature, and there is no way to modify them by adjusting any physical parameter. You cannot carve the steps differently. You cannot stand between steps. The discreteness is absolute.

Where the analogy holds, though, it holds well. Just as you cannot climb half a step on a staircase, an electron cannot absorb half a photon. Just as each step on a staircase has a specific height, each quantum transition corresponds to a specific energy difference. And just as the total height you ascend or descend is the sum of the individual steps you take, the energy an atom absorbs or emits is always an exact multiple of the fundamental quantum for that transition. The staircase picture is not just a metaphor. It is a geometric representation of what the mathematics actually says.

Diagram 2.8 – Staircase vs. Ramp: Quantized vs. Continuous Energy

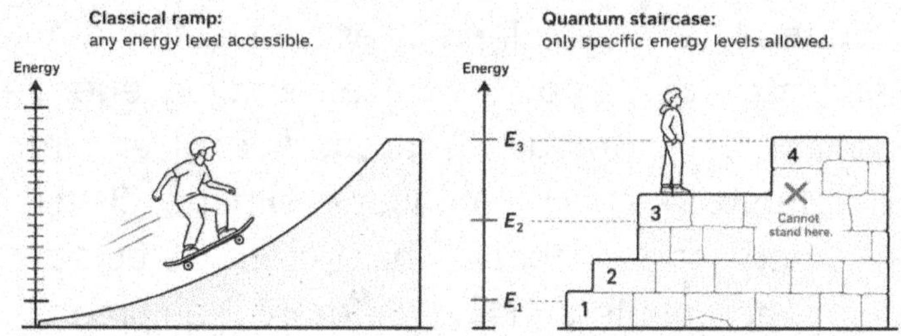


Quantum energy levels are not like stairs you could change, they are fixed by nature.

# Not a Theory, a Fact: Why Quantization Is Experimentally Necessary

At this point in the story, it is worth stepping back and asking a harder question. Could quantization be just a useful mathematical trick? Could there be some underlying continuous theory that produces the same results without truly discrete packets? The answer, based on a century of increasingly precise experiments, is no. Quantization is not a bookkeeping convenience. It is a feature of reality.

## Evidence from Multiple Independent Directions

Consider how many independent experiments all point to the same conclusion. The blackbody spectrum requires discrete oscillator energies. The photoelectric effect requires discrete photon energies. Atomic emission spectra require discrete electron energy levels. Experiments on the heat capacity of solids at low temperatures, problems that baffled classical physics entirely, are explained by assuming that atomic vibrations are quantized. Chemical bonding energies are quantized. The magnetic properties of electrons involve quantized spin, a concept with no classical analog. Every one of these phenomena, studied by different experimenters using different equipment in different countries, converges on the same conclusion: energy exchange at the atomic scale is discrete.

A particularly clean demonstration comes from the Franck-Hertz experiment, carried out in 1914. James Franck and Gustav Hertz sent electrons through mercury vapor and measured the electrons' energy loss in collisions with mercury atoms. If energy were continuous, the electrons should be able to lose any amount of energy in a collision, giving up a little energy to each mercury atom they encountered. What Franck and Hertz observed was completely different. The electrons lost energy only in fixed chunks corresponding to the energy difference between the mercury atom's ground state and its first excited level. Below that threshold, the electrons passed through the vapor without losing any energy: the mercury atoms refused to accept partial payments. The results provided direct, unambiguous proof that atomic energy levels were quantized and that energy transitions occurred in discrete steps.

Diagram 2.9 – The Franck-Hertz Experiment: Atoms Refuse Partial Payments

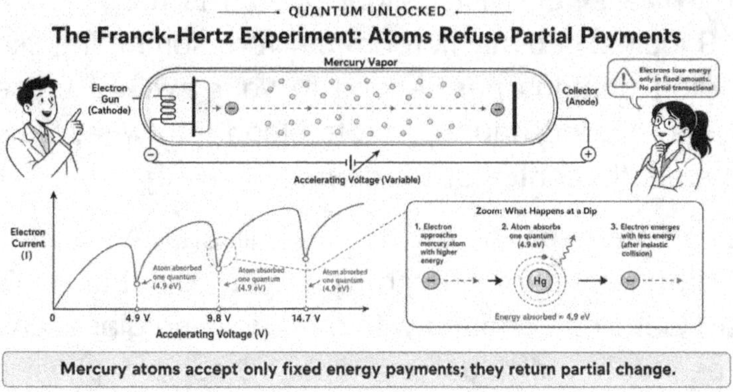


This point, that atoms refuse partial payments, is one of the most important ideas in the chapter. It is easy to think of quantization as a convenient approximation: energy comes in such tiny packets that they seem continuous, the way money comes in cents but seems continuous for large purchases. The Franck-Hertz experiment shows that this picture is wrong. An atom does not accumulate partial payments and release them when the total reaches the threshold. It simply does not accept anything below the minimum. This is fundamentally different from behavior in a continuous system, and it has measurable, dramatic consequences.

## Planck's Constant: The Grain Size of Reality

The reason quantization is invisible in daily life is that Planck's constant h is so small. One joule-second is the energy of a large apple dropped from a height of one meter, times one second. Planck's constant is about 6.6 times ten to the minus thirty-four of that unit. The energy of a single visible-light photon is roughly 3 × 10^-19 joules. A dimly lit room delivers something like ten billion photons per second to your eye. At those numbers, the individual packets blend into what feels like a smooth, continuous stream.

Think of the grain of a photograph. A printed photograph appears smooth when you hold it at arm's length. Get close enough with a magnifying glass, and you can see individual ink dots. Classical physics viewed the photograph from an arm's length away. Quantum physics has the magnifying glass pointed at individual dots. The grain is real. It was always there. We just lacked the instruments to see it until the right experiments were designed.

## The Quantum in Your Pocket: LED Lighting

Understanding quantization is not only an exercise in the history of physics. It is the operating principle behind technologies that are part of daily life, from the displays on our phones to the lights in our homes.

### How an LED Works

A light-emitting diode, or LED, is a device that produces light through a deliberate, engineered quantum transition. Inside an LED is a semiconductor material, typically a compound of gallium, indium, and nitrogen. When a voltage is applied, electrons are pushed into a region where they occupy high-energy levels. They then fall to lower energy levels, and each transition to a lower level releases a photon. The energy of that photon, and therefore its color, is determined by the energy gap in the semiconductor material.

LED engineers choose the semiconductor compound specifically to set the energy gap at the value that produces the desired color. A particular indium-to-gallium-to-nitrogen ratio yields a gap that produces blue photons. Coating the LED chip with a phosphor layer that absorbs blue photons and re-emits a broader spectrum produces what we perceive as white light. Every white LED bulb in your ceiling is running this two-step quantum process: first, generate blue photons through an engineered transition, then convert some of them to a broader spectrum via fluorescence.

The efficiency of an LED is a direct consequence of the sharpness of the quantum transition. Because electrons fall to a specific energy level rather than losing energy in an uncontrolled way, a high fraction of the electrical energy input is converted into photons of exactly the desired frequency. Traditional incandescent bulbs heat a filament until it glows. Because they rely on thermal radiation from a hot object, they emit energy across a broad spectrum: most of the energy goes into infrared photons that our eyes cannot see, producing heat rather than light. An LED wastes very little energy that way. The precision of the quantum transition is the source of its efficiency.

Diagram 2.10 – Inside an LED: Engineered Quantum Transitions

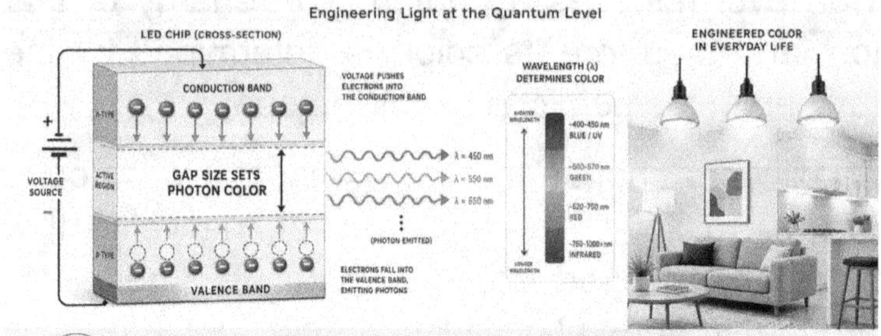


In an LED, the color of light is determined by a gap that engineers design at the quantum level.

## From Neon Signs to LED Screens

The path from Planck's 1900 formula to the LED screen in your hand took about a century of physics and engineering. The neon signs of the 1920s exploited natural atomic emission spectra by choosing gases whose spectral lines produced appealing colors. Engineers had no control over the energy levels; they chose from the palette that nature provided. LED technology, developed primarily in the 1960s through the 1990s, reached a new level: engineering the quantum levels themselves by selecting semiconductor compounds and layer thicknesses. Modern OLED displays take this further, using organic molecules whose quantum transitions can be tuned to produce precise RGB colors with extraordinary efficiency. In each generation, the underlying physics is the same. Energy transitions are quantized, photon energy is set by frequency, and the color of light is a direct read-out of the energy gap crossed.

This chain from basic experiment to transformative technology is the recurring theme of this book. Quantum mechanics was not invented for its practical applications. It was forced on physicists by experimental results that refused to fit the classical picture. The technologies followed are those that operate at the level where the real rules apply; once you understand the real rules, you can engineer at that level. Understanding quantization is not just historically important. It is the key to understanding why the machines you depend on every day work as well as they do.

## Bohr's Model: Triumphs and Limits

Bohr's model of the atom was a milestone. It explained the hydrogen spectrum, introduced the concept of quantized electron orbits, and gave physicists a concrete picture for reasoning. But it also had serious limitations that were visible almost immediately, and acknowledging those limits is part of doing honest science.

### Successes: Hydrogen and Hydrogen-Like Ions

The model worked extremely well for hydrogen, which has only one electron. It also worked for hydrogen-like ions: helium with one electron removed, lithium with two electrons removed, and similar single-electron systems. For all of these, the predicted spectral lines matched the experimental data to the precision of the available measurements at the time. The model also correctly predicted that there would be spectral lines in the ultraviolet and infrared, in series that had not yet been fully cataloged, and subsequent measurements confirmed those predictions.

## Failures: Multi-Electron Atoms and Beyond

The moment you tried to apply the Bohr model to helium, which has two electrons, it failed. The predicted spectral lines did not match the observations. The model provided no way to calculate spectral line intensities, only their frequencies. It could not explain why some transitions were much more probable than others. It gave no account of chemical bonding or of the periodic table. And it rested on a set of postulates, electrons in specific orbits without radiating, that had no theoretical justification within any known framework.

These failures were not embarrassments to hide. They were signposts pointing toward a deeper theory. The Bohr model was, in retrospect, a stepping stone: right enough to confirm that energy levels were real, but not detailed enough to describe atoms fully. The full theory of quantum mechanics emerged in 1925 and 1926 through the work of Werner Heisenberg, Erwin Schrödinger, and others. It replaced Bohr's circular orbits with a more abstract framework involving wave functions and probability distributions. But it retained the core insight: electron energies in atoms are quantized, and transitions between levels produce or absorb photons whose energies match the gaps. The river remains rainy.

Diagram 2.11 – Bohr's Model: Triumphs and the Path Beyond

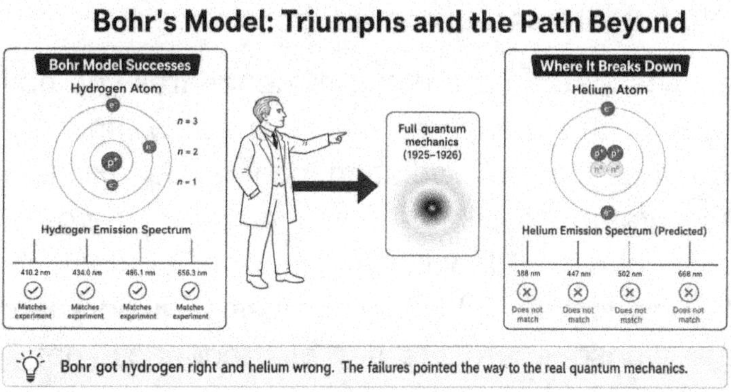


## Putting It All Together: The Investigator's Summary

Return to the opening image: water flowing from a tap. The flow is smooth because each water molecule is too small to see, and trillions of them pass through the tap each second. At human scales, the discrete molecular nature of water is invisible. But if you could shrink down to the size of a single water molecule, the flow would not look smooth at all. You would see individual molecules, discrete and separate, arriving one at a time.

Energy is like that, but more so. The discreteness is not just a matter of scale. It is a fundamental feature of energy exchange at the atomic level. An atom absorbing light does not slowly accumulate energy, as a battery is charged through a wire. It remains unchanged until a photon with exactly the right energy arrives, at which point it absorbs it entirely and jumps to the next level. The exchange is instantaneous and total. No partial transactions are possible.

As an investigator, here is the evidence trail you have followed in this chapter. The blackbody spectrum showed that thermal radiation cannot be explained without atomic oscillators exchanging energy in discrete chunks (Planck, 1900). The photoelectric effect showed that light itself arrives in fixed-energy packets whose energy is set by frequency, not by intensity (Einstein, 1905). Atomic emission spectra showed that atoms of each element have a fixed set of allowed energy levels and that photons are emitted only at frequencies corresponding to transitions between these levels. The Bohr model quantified those levels for hydrogen and matched the spectrum precisely (Bohr, 1913). The Franck-Hertz experiment directly confirmed that atoms absorb energy only in discrete units and refuse partial payments (Franck and Hertz, 1914). Every thread in the evidence leads to the same place. Energy at the atomic level is quantized. It is not continuous.

The technologies we examined, neon signs, fireworks, LEDs, and OLED displays, are not accidental applications of quantum physics. They work by exploiting precisely the fact that quantum levels fix photon energies. The engineer who designs a blue

LED chooses the spacing between quantized energy levels. The chemist who selects strontium for a red firework is choosing a color based on the element's atomic spectrum. The physicist who built the first neon sign was harnessing, without fully understanding, the same quantum phenomena that Planck and Einstein were working to explain at the same time.

One more idea is worth holding before moving forward. Quantum discreteness is not just a property of exotic laboratory situations. It is why atoms are stable. Without quantized energy levels, there is no ground state for the electron to occupy, and no reason for it to stop spiraling into the nucleus. The minimum energy level acts as a floor. The electron cannot fall below the first step. Quantization is not a complication added onto an otherwise sensible universe. It is the reason solid matter holds together. The chair you are sitting on exists because energy levels are discrete.

## Takeaway

A quantum is not jargon. It is a physical reality: an indivisible packet of energy whose size is set by frequency and Planck's constant. Light arrives in photons, each carrying $E = hf$. Atoms have fixed energy levels and can only absorb or emit photons at frequencies corresponding to the gaps between those levels. This is not a theoretical convenience. It is what experiment after experiment, from blackbody radiation to the photoelectric effect to the Franck-Hertz confirmation, has demonstrated beyond any reasonable doubt. Quantization is the reason atoms are stable, the reason each element has a unique spectral fingerprint, and the reason the light from a neon sign is that particular shade of orange-red and not some other color. When you see a lit LED bulb, a neon sign glowing in a window, or a burst of color in a firework, you are watching discrete quantum transitions in action, atoms trading energy in exact, indivisible packets, one photon at a time. The river was always raining. We needed the right experiments to see the drops.

# 3 The Double Agent: How Light Became Both a Wave and a Particle at Once

Picture yourself standing at the edge of a quiet pond. You drop a single stone into the center. Ripples fan outward in perfect circles. Now imagine dropping two stones at the same moment, a meter apart. Two sets of ripples spread out, and wherever the two sets meet, something interesting happens. In some places, the water rises higher than either stone alone could produce. In other places, the water goes completely flat, as though neither stone had fallen at all. This is not a trick of the light or a quirk of this particular pond. It is the fundamental behavior of waves, and it has been understood for centuries.

Now imagine performing that same experiment not with stones and water but with single particles of light shot one at a time through two narrow openings in a barrier. There is no pond. There is no second particle. There is only one particle, a single photon, traveling alone. And yet, over time, a pattern builds on a screen behind the barrier that looks exactly like the overlapping ripples in the pond. The pattern that emerges is the footprint of a wave interfering with itself. Something that traveled as a single particle produced the unmistakable signature of a wave.

This is the double-slit experiment. Richard Feynman, the Nobel Prize-winning physicist, called it the central mystery of quantum mechanics. He was not exaggerating. Every strange feature of the quantum world, every counterintuitive result that separates quantum physics from everyday experience, is contained in this one experiment. Your job as an investigator is to follow the evidence step by step, resist the temptation to reach for an explanation too early, and let the data tell its own story.

## Why It Matters

Understanding the double-slit experiment is not merely an exercise in appreciating a clever laboratory trick. It is the foundation for understanding what matter and energy actually are at the smallest scales. The computers and smartphones that run modern life are built from transistors whose behavior depends on quantum effects. Medical imaging technology, precision atomic clocks, and the coming generation of quantum computers all rest on the principles you are about to examine. Before any of those technologies can make sense, wave-particle duality must make sense. And wave-particle duality makes sense only when you understand what happened in the double-slit experiment and why it cannot be dismissed or explained away.

## Waves: The First Suspect

To understand what makes the double-slit result so startling, you first need to understand waves. Not the mathematics of waves, but the physical intuition behind what a wave actually does.

A wave is a disturbance that travels. When you pluck a guitar string, you set the string vibrating, and that vibration pushes the surrounding air. The air pressure rises and falls in a repeating pattern that spreads outward in all directions. Your eardrum picks up that oscillating pressure, your brain interprets it as sound, and you hear a note. The string itself did not travel across the room. The disturbance traveled. That distinction, energy and information moving through a medium without the medium itself being transported, is what makes something a wave.

Waves have several properties that particles do not. They have a wavelength, which is the distance between successive crests. They have a frequency, which is the number of crests that pass a fixed point per second. And they have amplitude, which is the height of each crest. These properties are measurable and predictable, and they combine in a specific way when two waves overlap. That combination is called interference.

Diagram 3.1 – Wave Anatomy: Wavelength, Amplitude, and Frequency

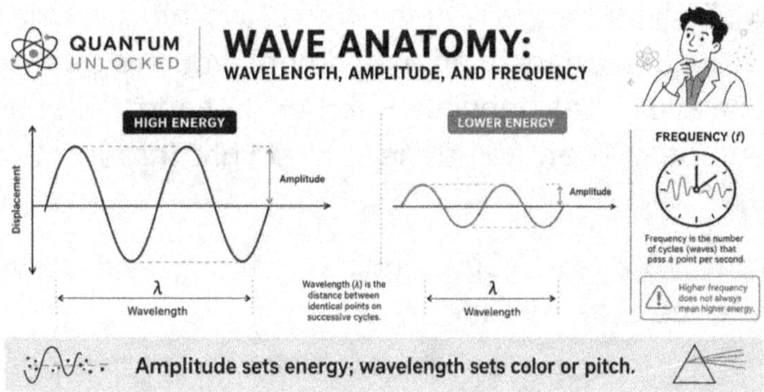


## Constructive Interference: When Waves Add Together

When two waves meet crest to crest, the result is a combined crest taller than either wave alone. The two waves reinforce each other. This is constructive interference. Where it occurs, the signal is amplified: water rises higher, sound gets louder, light grows brighter. Concert hall designers spend enormous effort shaping rooms so that sound waves arrive at every seat in phase, constructively combining to produce a rich, full sound rather than dead zones of cancellation.

## Destructive Interference: When Waves Cancel

The reverse situation is equally important. When the crest of one wave meets the trough of another, the two partially or completely cancel. At the exact point of overlap, the medium goes still. This is destructive interference. Crucially, interference is a behavior unique to waves. Two particles do not interfere. If you fire two bullets at a wall, they make two holes. When interference appears in an experiment, you are dealing with a wave. That principle seemed beyond question for centuries. Then someone shone light through two slits.

Diagram 3.2 – Constructive and Destructive Interference Side by Side

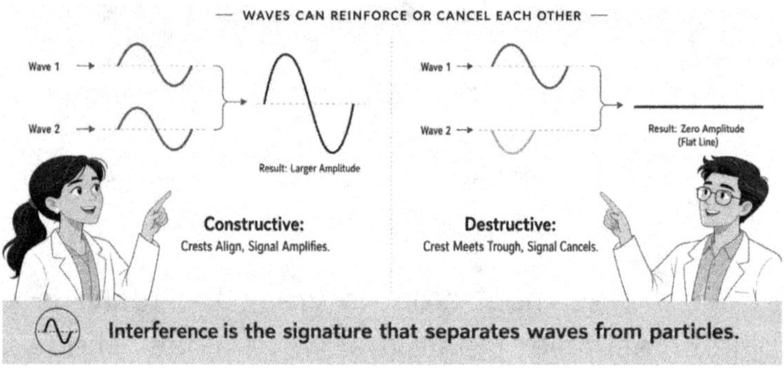


## Thomas Young & Light Experiment That Changed Everything

In 1801, an English physician and polymath named Thomas Young set out to settle a long-running argument. Isaac Newton had proposed, more than a century earlier, that light was made of particles, which he called corpuscles. The alternative view, championed by Christiaan Huygens, held that light was a wave. Newton's authority was so immense that the corpuscle theory dominated for decades, even though it struggled to explain certain optical phenomena. Young decided to let the light itself decide.

His experiment was elegant in its simplicity. He allowed a beam of sunlight through a single small hole, producing a coherent source, and let the light fall on a second card with two narrow slits close together. Behind the second card, he placed a screen. If Newton was right, he expected two bright bands, the way bullets fired through two holes leave two marks on a wall. What he saw instead was a series of alternating bright and dark bands spread across the screen. Not two bands. Many bands, fanning out symmetrically from the center, exactly the pattern you would expect if light waves from the two slits were interfering with each other, constructively at the bright bands and destructively at the dark ones.

Young had produced the unmistakable fingerprint of light interference. The experiment appeared to settle the question definitively. Light was a wave. The corpuscle theory was abandoned.

Diagram 3.3 – Young's Double-Slit Setup: From Single Source to Interference Pattern

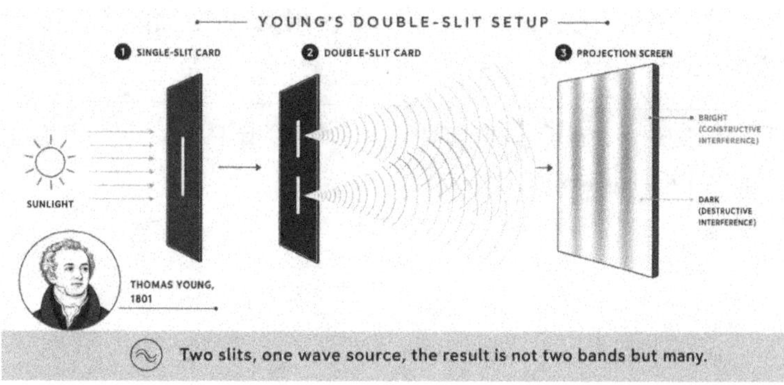

## Why the Pattern Appears Where It Does

The geometry of the interference pattern follows directly from the wave properties of light. Imagine standing at a point on the screen that is the same distance from both slits. Light waves from the two slits travel the same distance to reach you. They arrive in phase, meaning the crests align. Constructive interference occurs, producing a bright band at the center of the screen.

Now move slightly to one side. The distances from the two slits are no longer equal. If the path difference is exactly half a wavelength, a crest from one slit arrives at the same moment as a trough from the other, and you see a dark band. If the path difference is one full wavelength, crest meets crest again, and a new bright band appears. The alternating pattern repeats as you move across the screen. This is precise, calculable, and repeatable. Young used this relationship in reverse, measuring band spacing to determine the wavelength of light for the first time. The wave model was not just consistent with the data; it was quantitatively accurate.

Diagram 3.4 – Path Difference Geometry: Why Bright and Dark Bands Alternate

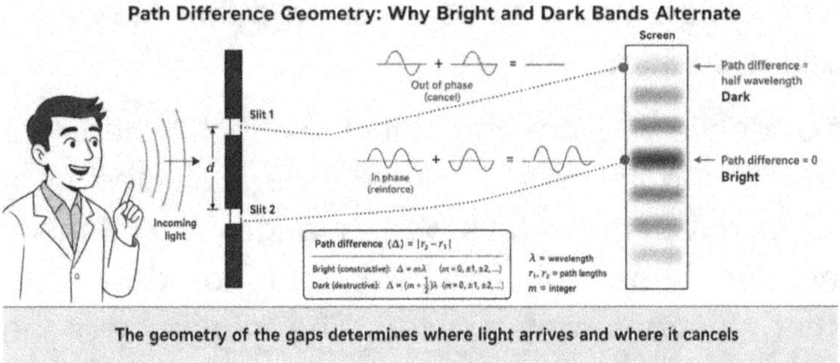


## Light as a Wave: A Century of Confidence

Young's experiment was followed by decades of reinforcing evidence. In the 1860s, the Scottish physicist James Clerk Maxwell produced a complete mathematical description of electricity and magnetism, and discovered something remarkable in the equations: they predicted the existence of waves made of oscillating electric and magnetic fields, traveling through space at a specific speed. When he calculated the speed, it exactly matched the measured speed of light. Light was an electromagnetic wave, a self-sustaining oscillation of electric and magnetic fields propagating through space.

Maxwell's theory was a triumph of 19th-century physics. It unified optics, electricity, and magnetism into a single coherent framework. Light was a wave. That was settled. For nearly a century after Young's experiment, the wave nature of light was not questioned. And then, at the very beginning of the 20th century, experimental evidence arrived that Maxwell's equations could not explain.

That evidence was the photoelectric effect. When light shines on certain metals, electrons are ejected from the surface. But the details were strange. The intensity of the light, how bright it was, did not determine whether electrons were ejected. The frequency of the light, its color, did. Dim blue light ejected electrons. Bright red light did not, no matter how intense. And when electrons were ejected, each one had the same energy, determined entirely by the frequency of the light.

This behavior made no sense if light was a continuous wave. A wave's energy is carried by its amplitude, its brightness. A more intense wave should always be able to knock out electrons. But the data showed that light arrived in discrete packets, each carrying energy proportional to its frequency. If a single packet lacked sufficient energy, no amount of additional packets would help.

In 1905, Albert Einstein proposed the explanation. Light also behaved as a stream of particles, each carrying a fixed quantum of energy. These particles were later named photons. Einstein won the Nobel Prize for this work in 1921, not for relativity, but for explaining the photoelectric effect. Light was both a wave and a particle. The double agent had been identified.

Diagram 3.5 – The Photoelectric Effect: Why Wave Theory Failed

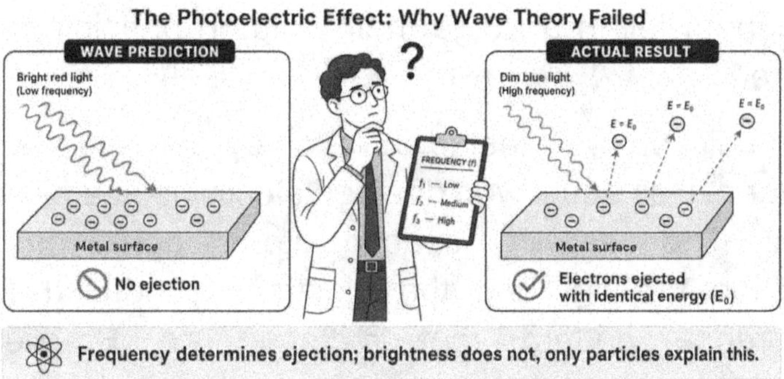


## The Experiment Repeated With Electrons

By the 1920s, the quantum revolution was reshaping physics. Planck had quantized energy. Einstein had quantized light. Niels Bohr had proposed a model of the atom in which electrons occupied discrete energy levels. And then, in 1924, Louis de Broglie, a French physicist writing his doctoral thesis, made a radical proposal that took the logic one step further.

If light, which everyone agreed was a wave, could also behave as a particle, perhaps the reverse was also true. Perhaps particles, things that everyone agreed were localized objects with mass and position, could also behave as waves. De Broglie proposed a formula relating a particle's momentum to its wavelength. The more momentum a particle carried, the shorter its wavelength. For a baseball, the implied wavelength was absurdly small, far smaller than any measurable scale. For an electron, which has very little mass and moves at relatively modest speeds, the implied wavelength fell in a range that could, in principle, be detected.

The prediction was tested in 1927 by Davisson and Germer at Bell Labs, who directed electrons at a nickel crystal. The regularly spaced atoms acted as a diffraction grating, and the electrons produced an interference pattern that matched de Broglie's predictions exactly.

The double-slit experiment with electrons came next. The setup was, in principle, identical to Young's light experiment. The electrons were fired at a double-slit barrier and detected on a screen behind it. Over time, the dots of individual electron impacts accumulated. And the pattern that appeared was not two bands. It was the same alternating bright-and-dark interference pattern that Young had seen with light. Then the experiment was made stranger still.

Diagram 3.6 – The Double-Slit Experiment With Electrons: The Impossible Pattern

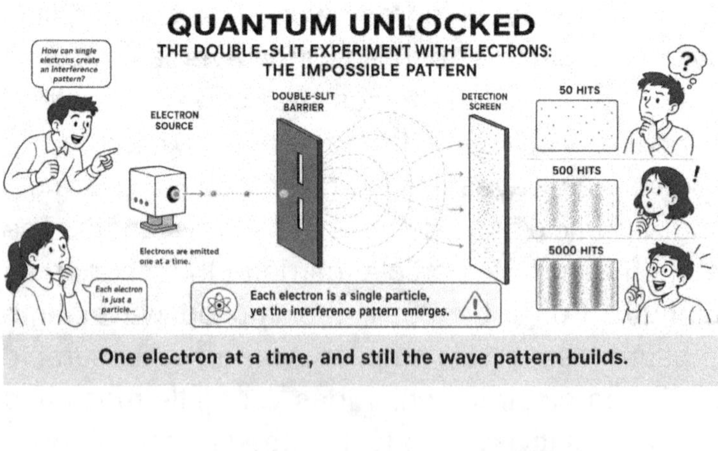

## One Electron at a Time

A natural objection runs as follows: the electrons must be interfering with each other, flying through in a crowd. This would be a classical explanation, and it would be wrong. The experiment was redesigned to eliminate that possibility. Electrons were emitted one at a time, with long enough intervals between them that no two were in flight simultaneously. Each electron traveled alone, in complete isolation, from source to screen.

The pattern that accumulated was identical. Each electron hit the screen at a single point, as if it were a particle. But over thousands of such single-particle impacts, the points accumulated into the interference pattern, bright bands where many hits clustered, dark bands where almost none appeared. Something about each electron's journey through the apparatus produced a probability distribution that mimicked a wave.

This result led Feynman to call the double-slit experiment the central mystery. You cannot say the electron went through the left slit. You cannot say it went through the right slit. If you assume it went through one or the other, you predict two bands. The experiment gives many bands. The mathematics requires you to treat the electron as a wave passing through both slits simultaneously and interfering with itself. But when it hits the screen, it hits as a particle at a single point.

The electron is not secretly a particle we cannot track. It is not secretly a wave that looks like a particle when detected. It behaves like a wave when propagating and like a particle when detected, and these two behaviors cannot be reduced to each other.

Diagram 3.7 – Single-Electron Build-Up: From Random Dots to Interference Pattern

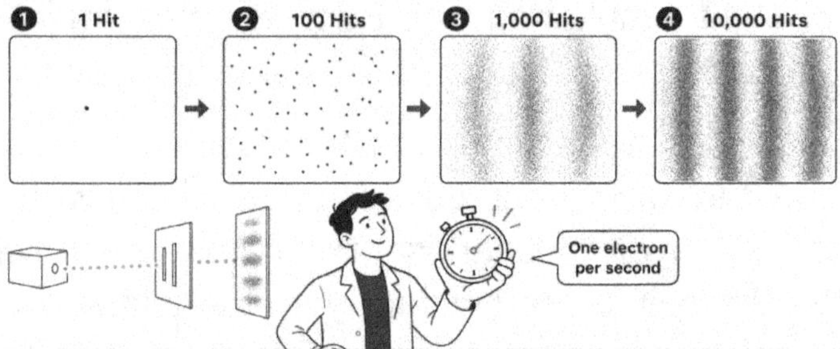


Thousands of single-particle events, yet a wave pattern emerges.

## The Ripples in the Pond, Revisited

Return for a moment to the pond analogy from the opening of this chapter. Ripples in a pond interfering with each other is a vivid and accurate picture of wave interference, and it captures the essential geometry of the double-slit result. The two slits act like the two points at which stones enter the water. The interference pattern on the screen is the pattern of high and low water that you would see spreading across the pond.

This analogy is useful and should be held onto. But it must also be corrected, because it fails in one critical way. The pond waves are water waves. There is a physical medium, actual molecules of water, oscillating up and down. When we say an electron behaves like a wave, we cannot point to a medium. There is no electron-water. The wave is not a wave in any physical substance. It is a wave of probability, a mathematical description of where the electron is likely to be found. The wave crests are regions of high probability. The troughs are regions of low probability. The interference pattern is a pattern of probabilities, not of any physical substance.

This distinction is important and will become central in later chapters. For now, notice that the distinction does not make the phenomenon less real. The interference pattern produced by single electrons is measured, reproduced, and predicted with precision. Whatever the electron's wave nature consists of, it has real, measurable consequences.

## Wave-Particle Duality: What the Term Actually Means

The phrase "wave-particle duality" is used so often and in so many different contexts that it has accumulated layers of misunderstanding. Before going further, it is worth being precise about what the term means and what it does not mean.

Wave-particle duality does not mean that electrons sometimes look like waves and sometimes look like particles depending on some changeable property of the electron, the way a person might look different in different lighting. It does not mean that electrons are really particles, but rather that they have a wave attached to them, the way a boat has a wake. It does not mean that we are ignorant of the electron's true nature or that a better theory will eventually reveal it to be just one thing.

What wave-particle duality means is this: the behavior of quantum objects cannot be fully captured by either the classical concept of a wave or the classical concept of a particle. In situations that test wave behavior, quantum objects exhibit wave-like behavior. In situations that test particle behavior, quantum objects exhibit particle-like behavior. These two aspects are complementary: each is revealed in a different experimental context, and no single experiment can reveal both simultaneously.

Niels Bohr formalized this as the principle of complementarity. The wave and particle aspects of a quantum object are not contradictory; they are complementary descriptions, each valid in its own domain, each incomplete on its own. Together, they describe something the classical world has no category for.

Diagram 3.8 – Wave-Particle Duality: Two Faces of the Same Object

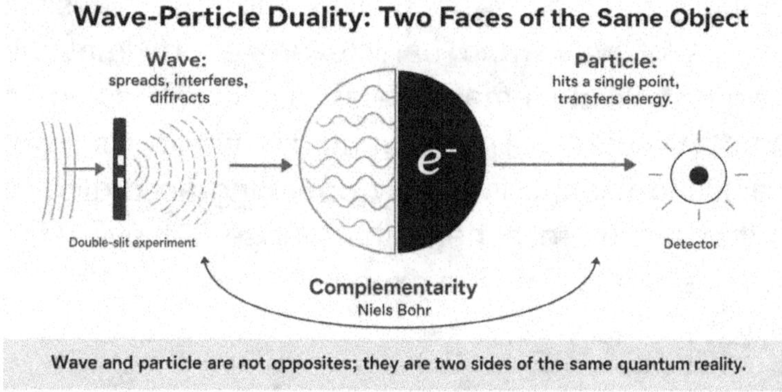


## Why Classical Concepts Fall Short

The classical world gave us the concept of a particle: a localized object with a definite position and momentum at every moment. The classical world also gave us the concept of a wave: a spread-out disturbance that propagates through a medium or field. These are two fundamentally different categories. A billiard ball is a particle. A sound wave is a wave. There is no overlap.

Quantum mechanics reveals that this clean division is a feature of the scale at which human beings happen to live, not a feature of nature itself. At the atomic scale, the categories blur. Electrons, protons, neutrons, and photons all exhibit both behaviors. The properties we think of as exclusively belonging to waves, interference, and diffraction, are observed in particle experiments. The properties we think of as exclusively belonging to particles, localized energy deposits, and discrete detection events, are observed in wave contexts.

A useful way to think about it is this: the classical categories of wave and particle are approximations that work well at human scales but break down at quantum scales. The quantum world is not contradicting itself; it is simply operating according to rules that are more general than those our everyday experience prepared us to expect.

## The Role of the Detector: Watching Changes Everything

If the previous sections raised a question in your mind, you are a good investigator. The question is this: if we cannot say which slit the electron went through, what would happen if we set up a detector at the slits to find out? What if we watched?

This is not a thought experiment. It has been done, in many variations, with carefully designed apparatus. The result is one of the most consequential and most discussed findings in the history of science. When a detector is placed at or near the slits to determine which slit each electron passes through, the interference pattern disappears. The screen shows two bands, exactly the pattern you would expect if the electrons were ordinary particles passing through one slit or the other.

Let that sink in. When no detector is present, electrons produce an interference pattern, the signature of wave behavior, as though each electron passed through both slits. When a detector is present, electrons produce a two-band pattern, the signature of particle behavior, as though each electron passed through exactly one slit. The act of detecting which slit the electron passed through destroys the interference.

This result holds regardless of how the detector is constructed, as long as it actually records which-way information. Physicists have tried extremely gentle detectors, detectors that interact with the electron as little as possible. If the detector records which slit is used, the interference vanishes. If the detector is switched off so that no which-way information is available, the interference returns. The pattern depends not on whether the detector physically disturbs the electron, but on whether which-way information exists anywhere in the universe.

Diagram 3.9 – Detector On vs. Detector Off: The Pattern Changes With Observation

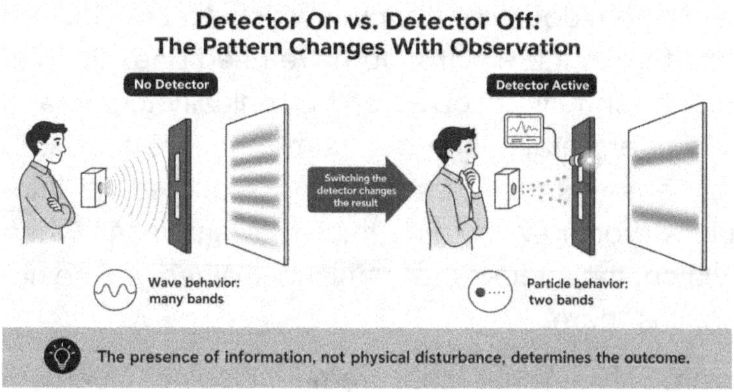

## What This Means for the Concept of Measurement

The detector result forces a profound revision in how we think about measurement. In classical physics, measuring something means finding out what was already there. A thermometer does not change the room's temperature; it reads a temperature that existed before the thermometer was inserted. Measurement is passive discovery.

In quantum mechanics, measurement is not passive discovery. The act of gathering which-way information does not reveal a pre-existing trajectory; it helps determine the outcome. Before measurement, the electron's wave nature predicts the interference pattern. After a which-way measurement, the electron's behavior collapses to particle behavior. The measurement does not discover which slit the electron used; it forces the electron to have used one slit. This has not been fully resolved philosophically today, and several interpretations of quantum mechanics exist to make sense of it. These will be explored in later chapters. For now, the key finding is empirical: when you watch, the interference vanishes. When you do not watch, it is there.

Diagram 3.10 – What Measurement Does in Quantum vs. Classical Physics

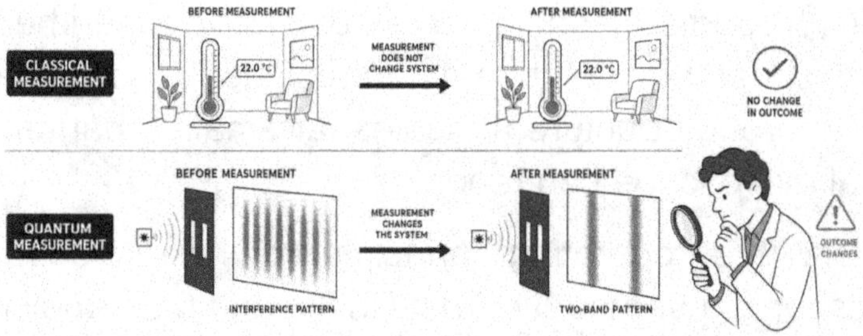


In the quantum world, looking changes what you see, not because of clumsiness, but by definition.

## Electron Microscopy: Putting Wave Nature to Work

The wave nature of electrons is not merely a philosophical puzzle. It is a tool, and a powerful one. To understand why, you need to know one basic fact about optical microscopes: their resolution, the smallest detail they can distinguish, is limited by the wavelength of the light they use. You cannot imagine an object smaller than the wavelength of your illuminating beam. Visible light has wavelengths between roughly 400 and 700 nanometers. That means an ordinary microscope cannot resolve anything smaller than a few hundred nanometers, which is large enough to see bacteria but too large to see individual molecules.

Electrons, however, have a wavelength that depends on their speed. An electron accelerated by a modest electric potential acquires a de Broglie wavelength that is thousands of times shorter than that of visible light, measured in fractions of a nanometer. If electrons have wave properties, and the double-slit experiment confirms that they do, then a beam of electrons can be used to probe structure at scales far smaller than any optical microscope can reach.

This is the operating principle of the electron microscope, invented by Ernst Ruska and Max Knoll in the early 1930s. Instead of glass lenses, an electron microscope uses magnetic fields to focus electron beams. Instead of visible light, it uses the quantum wave nature of electrons as its probe. The result is the ability to image individual atoms, to see the actual physical structure of materials at the scale of angstroms, one ten-billionth of a meter.

Today, electron microscopes are standard equipment in materials science, biology, chemistry, and nanotechnology. Every image you have seen of a virus particle, a microprocessor circuit, or a crystal's atomic lattice was likely captured using an instrument that exists because electrons have wave properties. The strange result of the double-slit experiment is not a philosophical curiosity; it is embedded in the infrastructure of modern science and technology. Quantum mechanics is the description of how the world actually works at the most fundamental level, and every technology that operates at small scales depends on the principles you are examining.

Diagram 3.11 – Electron Microscopy: Wave Nature as Scientific Instrument

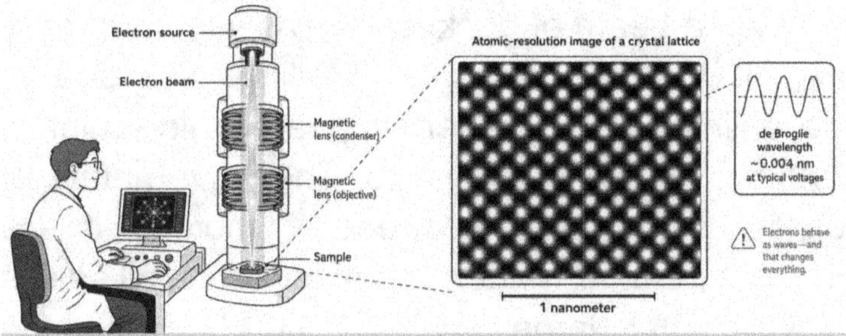


The same wave property that creates the interference pattern enables atomic-scale imaging.

## Why the Interference Pattern Cannot Be Explained Away

At this point in the investigation, it is worth pausing to address an instinct to be skeptical. You might wonder whether the interference pattern is somehow an artifact of the apparatus, a byproduct of the physical structure of the slits or the detector, rather than a genuine property of the electrons. You might wonder whether the electrons are really being disturbed by the slit edges in a way that happens to mimic interference. These are good scientific instincts, and they have been thoroughly explored.

Physicists have performed the double-slit experiment in countless variations, using different particles including protons, neutrons, entire atoms, and large molecules. Every time, the interference pattern appears when which-way information is absent and disappears when it is present. One particularly striking variation used buckminsterfullerene molecules, also known as buckyballs, which contain sixty carbon atoms each. When a beam of buckyballs was sent through a double slit in 1999 by Anton Zeilinger's group in Vienna, an interference pattern appeared. Molecules of sixty atoms, behaving as waves. The wave nature of matter is not special to electrons or photons. It is a general feature of all matter, prominent at small scales and invisible at large scales, only because the wavelengths become too tiny to detect.

The pattern cannot be explained away. It is not a measurement artifact, not a statistical quirk, not a theoretical abstraction. It is the actual behavior of matter when it is not being watched. The investigator who reaches this point in the evidence has no choice but to accept what the data says: something that we call a particle can travel through two openings simultaneously, interfere with itself, and then arrive as a single localized impact. Nature, at its smallest scales, is genuinely doing this.

## The Honest Reckoning

Many popular accounts of the double-slit experiment end here, suggesting that quantum mechanics is just weird and must be accepted without understanding. This book does not do that. The strangeness is real, but it is navigable. Understanding why the experiment yields the results it does, what the wave function represents, and what happens at the moment of detection requires careful, well-developed answers. The following chapters will systematically build those answers.

What you should carry forward is not vague wonder but a precise set of findings. Waves interfere; particles do not. The double-slit experiment with light produces interference, establishing the wave nature of light. The same experiment with electrons produces the identical interference, establishing that matter also has wave properties. When which-way information is available, the interference disappears. The experiment holds at all scales tested, from photons to large molecules. These are the facts of the investigation.

## Takeaway

You can now describe the double-slit experiment from memory and explain what the interference pattern means. Light, when directed through two slits, produces a pattern of bright and dark bands that is the unmistakable signature of wave interference. Electrons, sent through the same kind of barrier one at a time, produce the identical pattern, even though each electron arrives at the screen as a single localized impact. The pattern builds up over thousands of individual events, and it cannot be explained by electrons interacting with one another, because it persists even when electrons are sent one at a time, with no others in flight.

Wave-particle duality is the term for this behavior, but the term should be read carefully. It does not mean quantum objects flicker between two classical identities. It means that the classical categories of wave and particle are both insufficient to describe quantum objects, which exhibit the distinctive features of each in the appropriate experimental context. The wave and particle aspects are complementary: each is revealed by a different kind of experiment, and no experiment can fully reveal both at once.

The detector's role is not a secondary detail. It is the central finding. When a detector is placed at the slits, and which-way information becomes available, the interference pattern vanishes, and the electrons behave as classical particles. When no which-way information is available, the interference pattern reappears. The act of gathering information, not any physical disturbance caused by the detector, is responsible for the change in behavior. This finding points to a deep connection between information and physical reality that will be explored throughout the rest of this book.

The wave nature of electrons is not just a laboratory curiosity. It is a practical tool. Electron microscopes exploit the short de Broglie wavelength of electrons to image structures at atomic resolution, producing the pictures of viruses, crystal lattices, and microchip circuits that are now familiar in science and engineering. Every time a researcher uses an electron microscope, they are relying on the same wave behavior demonstrated in the double-slit experiment.

The double-slit experiment is the gateway through which every reader of this book must pass. On the other side, the quantum world opens up: uncertainty principles, superposition, entanglement, and the deep question of what measurement actually means. All of those phenomena connect back to the findings in this chapter. The interference pattern is not a puzzle to solve and set aside; it is a signpost pointing toward a description of reality that is richer, stranger, and more precise than the classical world our intuitions were built to navigate.

# 4 The Fog of Certainty: What Heisenberg Actually Discovered

Picture yourself standing in a park on a bright afternoon. A hummingbird appears at a flowering shrub. You raise your camera. You want two things at once: a perfectly sharp image of the bird's position in the frame and a clear record of its motion. But you cannot have both. If you use a fast shutter speed to freeze the bird in a single spot, you capture its precise location. The blur that would have told you where it was heading disappears. If you slow the shutter to reveal the arc of its wings, the bird itself smears into a streak. You have traded one kind of information for another. This is a useful picture. It is also, in an important way, wrong. And understanding exactly how it is wrong is the first step toward grasping one of the deepest truths in all of physics.

Werner Heisenberg was twenty-five years old when he worked out the mathematics that would bear his name. He was recuperating from hay fever on the island of Helgoland in 1925, filling notebooks with a new kind of calculation that replaced the comfortable trajectories of classical physics with something far stranger: arrays of numbers, multiplied in sequences that did not commute. That non-commutativity was not a mathematical curiosity. It was a signal from nature itself. It said: there are pairs of properties in this world that do not merely resist being measured simultaneously. They do not both exist, with full precision, at the same moment. This chapter investigates what that actually means, why the camera analogy fails, and why atoms would collapse without this feature of reality.

## Why It Matters

The uncertainty principle is the most quoted and most misunderstood result in quantum physics. In popular culture, it has been used to justify everything from the observer effect in psychology to mystical claims about consciousness shaping reality. None of that is what Heisenberg found. What he actually found is both stranger and more specific: a hard, quantitative limit on how well certain pairs of physical properties can be defined simultaneously and not just measured. Defined. This is not a statement about cameras, electron microscopes, or scientists' clumsiness. It is a statement about the structure of reality. Understanding it correctly transforms your picture of atoms, energy, space, and why the universe is stable at all.

## The Camera Analogy and Its Limits

### What the Analogy Gets Right

The hummingbird photograph is a fine place to start because it captures the trade-off structure of the uncertainty principle. When you sharpen your measurement of one quantity, you blur your knowledge of a related one. In photography, this happens because you are choosing how to collect light over time. A short exposure captures light from a single frozen instant. A long exposure collects light over time. You cannot have both the temporal snapshot and the temporal history in the same image. The trade-off is real, physical, and unavoidable given what a photograph actually is.

Heisenberg's principle has the same shape: know position precisely, and momentum becomes uncertain; know momentum precisely, and position becomes uncertain. The trade-off is real, physical, and unavoidable given what a quantum particle actually is. So the analogy earns its place in introductory explanations. It prepares the mind for a genuine relation between two kinds of information.

Diagram 4.1 -- The Hummingbird Trade-off: Sharp Position vs. Blurred Motion

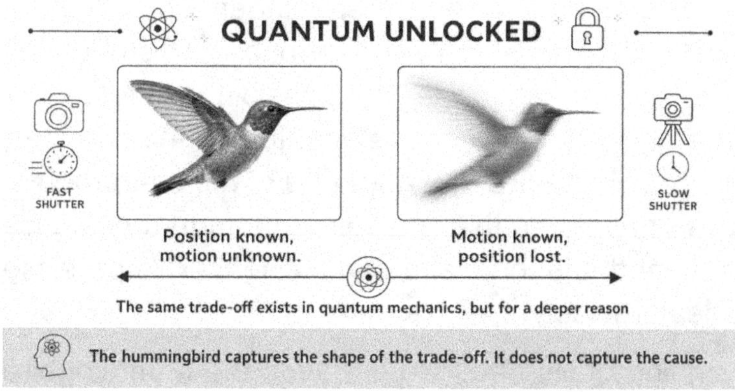

## Where the Analogy Breaks Down

Here is where the investigator must be careful. The hummingbird photograph involves a failure of technique. In principle, a fast enough shutter, a sensitive enough sensor, and enough light would let you capture both position and motion with arbitrary precision. The trade-off in photography is a technological limit, not a fundamental one. You are choosing how to collect information. The bird itself, at every moment, has a perfectly definite position and a perfectly definite velocity. You are just not capturing both at once.

Quantum particles are different in kind, not just in degree. An electron does not have a perfectly definite position and a perfectly definite momentum; we cannot record both simultaneously. Before measurement, those quantities are genuinely undefined, or more precisely, they exist only as probability distributions spread across a range of possible values. The uncertainty is ontological: it is a feature of the particle's actual state in nature, not a feature of our instruments. This is the claim that makes quantum mechanics strange, and it is the claim that Heisenberg's principle encodes precisely.

The analogy fails because it leaves the hummingbird intact as a classical object with definite properties. In quantum mechanics, the electron before measurement is not a tiny classical ball whose position we are blurring. It is a quantum object that has no single definite position or momentum. The measurement does not merely reveal a pre-existing value. It participates in producing one.

Diagram 4.2 -- Ontological vs. Epistemic Uncertainty

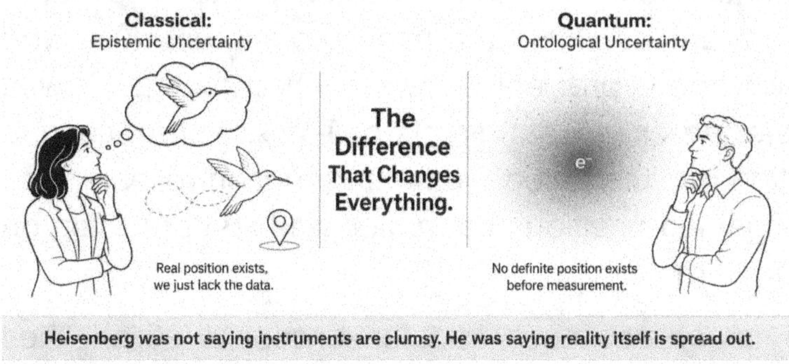


# Heisenberg's Gamma-Ray Microscope

## The Thought Experiment

Heisenberg himself worked through the principle using a thought experiment so specific that it has a name: the gamma-ray microscope. Imagine you want to measure an electron's position with extreme precision. To locate something very small, you need very short-wavelength light, because resolution is limited by wavelength. Visible light has wavelengths far too large to pin down an electron's position. You need gamma rays, which have wavelengths short enough to resolve atomic-scale positions.

So you illuminate the electron with a gamma-ray photon. The photon bounces off the electron and enters your microscope, telling you where the electron was. Position measured. Success. But here is the problem: gamma-ray photons carry a great deal of momentum. When such a photon strikes an electron, it kicks it. The collision transfers some of the photon's momentum to the electron. Because photons scatter at random angles within the microscope's aperture, you cannot know precisely how much momentum was transferred or in what direction. The act of measuring position has disrupted momentum uncontrollably. You have gained position information and lost precision on momentum.

This is where most popular accounts stop. They present the gamma-ray microscope as the whole story: measurement disturbs the system, and that disturbance is the uncertainty principle. But Heisenberg himself understood that this thought experiment reveals the principle rather than explaining it. The deeper question is: why does the act of measurement disturb the system at all? The answer lies in the wave nature of quantum objects.

Diagram 4.3 -- The Gamma-Ray Microscope: Illuminating an Electron

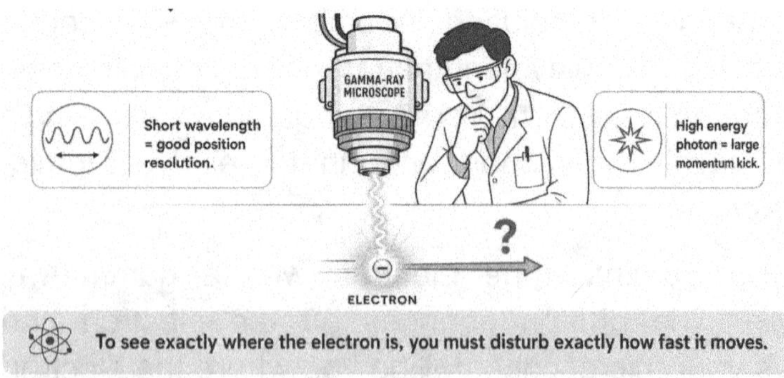

## Beyond the Disturbance Picture

The gamma-ray microscope makes the principle vivid, but there is a more fundamental source. Quantum particles behave like waves. A wave that is perfectly localized in space is not a simple, smooth oscillation. It is built from a superposition of many different wavelengths. In the language of quantum mechanics, wavelength encodes momentum: longer wavelength means lower momentum, shorter wavelength means higher momentum. A particle with a perfectly definite position corresponds to a wave that is sharply concentrated at one point in space. But such a wave requires contributions across an enormous range of wavelengths, which in turn implies an enormous range of momenta. You cannot build a sharp position spike from a single wavelength.

Conversely, a particle with perfectly definite momentum corresponds to a wave with a single wavelength, a pure sinusoidal oscillation that extends infinitely in both directions through space. Such a wave has no localization at all. Its position is entirely undefined. It is everywhere and nowhere with equal probability.

This is the heart of the matter. Position and momentum are not just difficult to measure simultaneously; they are also difficult to measure simultaneously. Their mathematical representations are related by a transformation that makes them fundamentally incompatible with simultaneous precision. The sharper one is, the broader the other must be. This relationship is built into the wave-like nature of quantum objects and cannot be engineered away, no matter how precise the instruments become.

Diagram 4.4 -- Wave Packets: Localization Requires Many Wavelengths

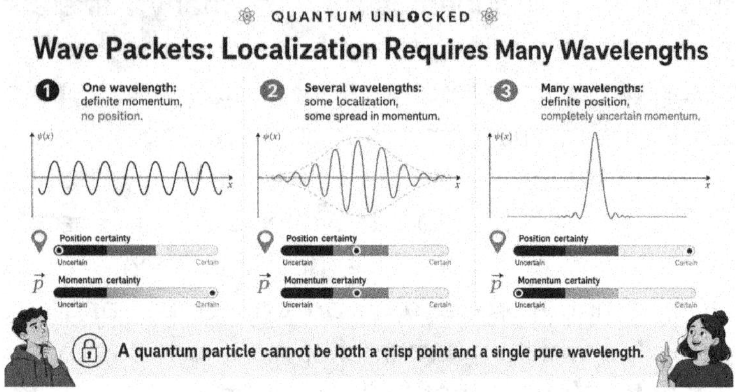


## What Uncertainty Actually Says

### The Principle Stated Plainly

Heisenberg's uncertainty principle, stated without mathematics but without distortion, says this: for any quantum particle, the product of the uncertainties in position and momentum cannot be made smaller than a fixed, irreducible quantity. Planck's constant sets that quantity, the same fundamental number that appears in the photoelectric effect and in every corner of quantum mechanics. Planck's constant is very small by everyday standards. This is why you do not notice quantum uncertainty when you catch a baseball. The ball's position and momentum are known to precision many orders of magnitude beyond the quantum limit. For electrons and protons, the limit becomes decisive.

The principle also applies to other pairs of so-called complementary variables. The most important second pair is energy and time. The uncertainty in the energy of a quantum state multiplied by the time that state persists cannot be smaller than that same irreducible quantity. A state that exists for a very short time necessarily has a large spread in energy. A state that persists for a very long time can have an extremely precise energy. This second form of the principle has consequences that are every bit as surprising as the position-momentum version.

Diagram 4.5 -- The Uncertainty Trade-off: Visualizing the Inverse Relationship

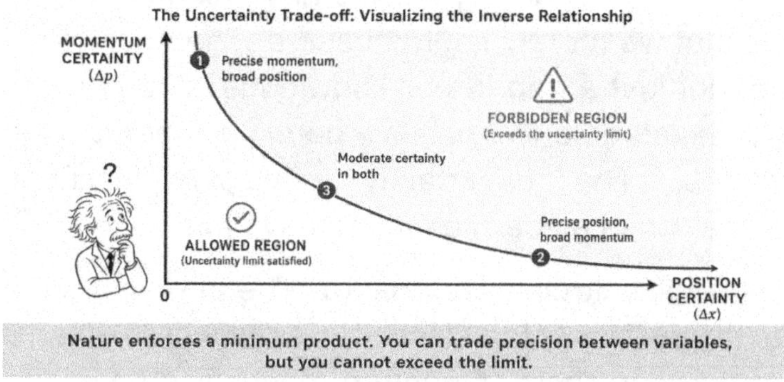


## Epistemic vs. Ontological: The Crucial Distinction

At this point, the investigator should press on a critical question. Is the uncertainty in quantum mechanics epistemic, meaning a limit on what we can know, or is it ontological, meaning a limit on what exists? These are genuinely different claims. An epistemic limit says the electron has a definite position and momentum right now; we cannot find both. An ontological limit says there is no fact of the matter about both simultaneously. The electron's position is not merely unknown; it is unknown. Before measurement, it has no single definite value.

This question was at the heart of the debate between Niels Bohr and Albert Einstein, which will be taken up more fully in later chapters. For now, the experimental record points firmly toward the ontological interpretation. Bell's theorem, proved in 1964 and tested exhaustively since the 1970s, rules out the possibility that quantum particles have hidden definite values that we merely fail to observe. The uncertainty is real, not just a gap in our knowledge. The particle, before measurement, inhabits a genuine superposition of states. The uncertainty principle describes superposition in a specific, quantitative form.

This is a difficult claim to absorb. The classical mind wants to say: surely the electron is somewhere, even if I do not know where. Quantum mechanics says: the question 'where is the electron right now?' has no answer when the electron is not being observed, because 'right now' requires a kind of sharp, definite state that quantum mechanics forbids.

# Energy and Time: The Second Pair

## A Different but Related Uncertainty

Most explanations of the uncertainty principle stop at the relationship between position and momentum. But the energy-time version carries consequences that reach further into everyday reality, or rather into the apparent emptiness of space. Consider the energy budget of a quantum system. If a particle exists in an energy state for a very long time, that state's energy is sharp and well-defined. The longer it lives, the more precisely its energy can be specified. But if that same state is extremely short-lived, the energy is correspondingly fuzzy. The state can have more or less energy than its average, and the shorter its lifetime, the wider the spread of energy.

This has a direct consequence for what appears to be space. According to quantum field theory, the vacuum is nothing. It is a quantum system with a ground state and, like all quantum systems, is subject to the uncertainty principle. In any tiny region of space over any tiny interval of time, the energy can fluctuate. Pairs of particles can briefly spring into existence, borrowing energy from that fluctuation, as long as they disappear before the principle demands repayment. These are called virtual particles.

Diagram 4.6 -- Energy-Time Uncertainty: Virtual Particles in the Vacuum

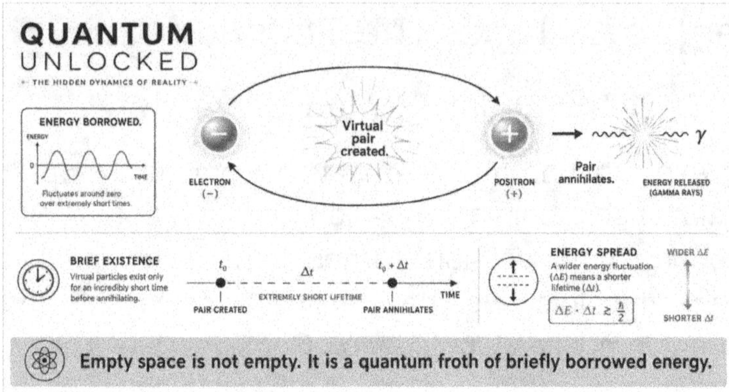


## Virtual Particles Are Not Hypothetical

The phrase 'virtual particles' sometimes sounds like a theoretical convenience, an accounting trick with no physical reality. In fact, the effects of virtual particles have been measured with extraordinary precision. The Casimir effect is one of the clearest demonstrations. Place two perfectly flat metal plates very close together in an otherwise empty vacuum. The plates restrict which wavelengths of virtual photons can exist between them. Wavelengths that do not fit between the plates are excluded from the interior. Outside the plates, all wavelengths can exist. This asymmetry creates a pressure difference. The energy difference between the outside and inside vacuum states pushes the plates together. The force has been measured in laboratory experiments and matches theoretical predictions to high precision. Space, regulated by the energy-time uncertainty principle, pushes two metal plates together.

The Lamb shift is a subtler example. Quantum electrodynamics predicts that virtual particles briefly surrounding a hydrogen atom should perturb the electron's energy levels in a tiny but specific way. Willis Lamb measured this shift in 1947. The result matched the theory. Virtual particles are not decoration. They are a quantitative prediction of the energy-time uncertainty principle, which has been experimentally confirmed.

Diagram 4.7 -- The Casimir Effect: Vacuum Energy Between Two Plates

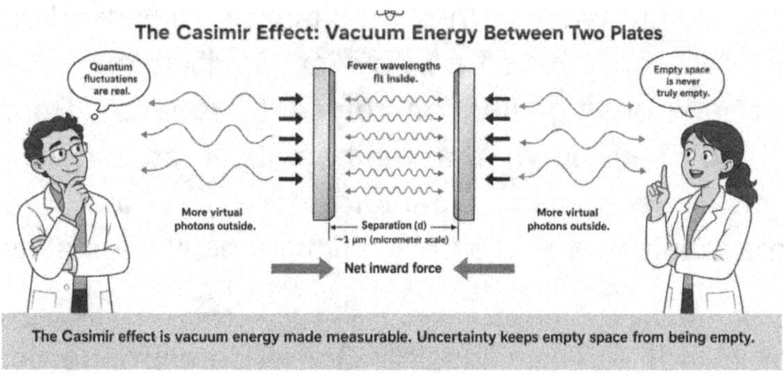


## Why Electrons Do Not Spiral Into the Nucleus

### The Classical Catastrophe

Before quantum mechanics, physicists faced a devastating puzzle about atomic structure. Rutherford had shown in 1911 that atoms consist of a tiny, dense, positively charged nucleus surrounded by orbiting electrons. But this picture was immediately catastrophic from the standpoint of classical electromagnetism. A charged particle moving in a curved path accelerates. An accelerating charged particle radiates electromagnetic energy. An electron orbiting a nucleus should therefore continuously shed energy as radiation and spiral inward, crashing into the nucleus in a tiny fraction of a second. Classical physics predicted that matter was unstable. Yet atoms persist. Something had to be wrong.

Bohr patched the problem in 1913 with a set of rules: electrons could only occupy certain allowed orbits and did not radiate while in those orbits. The rules worked numerically but offered no physical justification. Why should an orbiting electron refuse to radiate? Bohr could not say. It was a mathematical prescription without a mechanism. Quantum mechanics provided the mechanism, and the uncertainty principle lies at its heart.

Diagram 4.8 -- The Classical Spiral Catastrophe vs. Quantum Stability

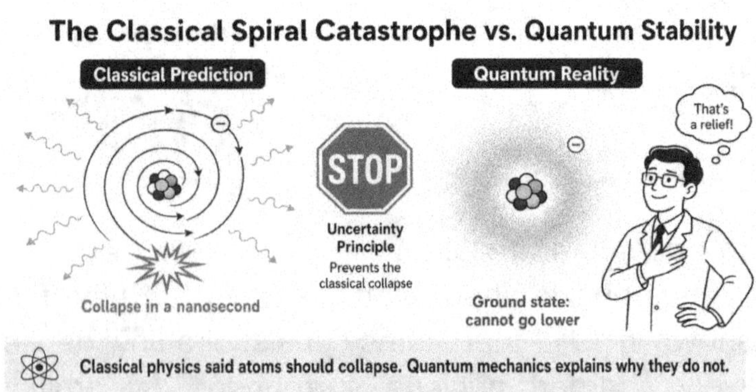

## How the Uncertainty Principle Stabilizes Atoms

The uncertainty principle prevents the classical collapse through a simple but powerful argument. Suppose the electron were confined to a region very close to the nucleus. A very small confinement means a very precise position. A very precise position means, by the uncertainty principle, a very large spread in momentum. A large spread in momentum indicates the electron has a large average kinetic energy. The more tightly you try to pin the electron against the nucleus, the more kinetic energy it necessarily possesses. At some point, the kinetic energy associated with tight confinement outweighs the electrical energy gained by being close to the nucleus, and the electron is energetically pushed outward. The balance between the electrostatic attraction that pulls the electron in and the quantum uncertainty that pushes it out defines the atom's ground state. That ground state is stable. The electron cannot fall further, not because of a rule imposed by Bohr, but because nature will not allow a state of simultaneously small position uncertainty and small kinetic energy.

This is one of the most satisfying applications of the uncertainty principle, because it explains a fact you can verify by looking at your own hand. Your hand is made of atoms. Those atoms have been stable for billions of years. They are stable precisely because quantum mechanics forbids the classical spiral. The uncertainty principle is not just a quirk of physics laboratories. It is the reason matter exists in the form that you are.

Diagram 4.9 -- How Quantum Uncertainty Stabilizes the Atom

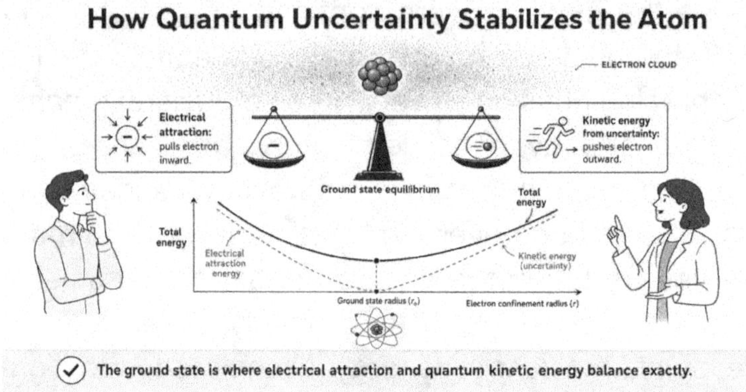


# Heisenberg's Matrix Mechanics and Wave Mechanics

## Two Routes to the Same Truth

Quantum mechanics was discovered twice in a single year, by two different people using two entirely different mathematical frameworks. Heisenberg, working in 1925, developed what became known as matrix mechanics. He represented physical quantities not as single numbers but as arrays of numbers arranged in grids, called matrices. The non-commutative multiplication of these matrices was what produced the uncertainty relation. If matrices represent position and momentum, and if those matrices do not commute, then the product of their uncertainties is bounded below by a fixed quantity. The uncertainty principle is not just consistent with matrix mechanics. It is built directly into the algebraic structure.

Erwin Schrodinger, working in 1926, arrived at a completely different formulation. He wrote a differential equation that describes how a wave function evolves in space and time. His equation was easier for many physicists to use because it resembled classical wave equations they already knew. Schrodinger's approach produced the same predictions as Heisenberg's but looked nothing like it. The wave nature of particles was central to Schrodinger's picture, while Heisenberg's matrices did not refer to waves at all.

The two approaches seemed irreconcilable at first. Heisenberg found Schrodinger's wave functions physically obscure. Schrodinger found Heisenberg's matrices algebraically off-putting. Within a year, Paul Dirac and others proved that the two formulations were mathematically equivalent: any problem solvable in one framework could be solved in the other, and they would yield identical predictions.

Diagram 4.10 -- Two Roads to Quantum Mechanics: Matrix vs. Wave

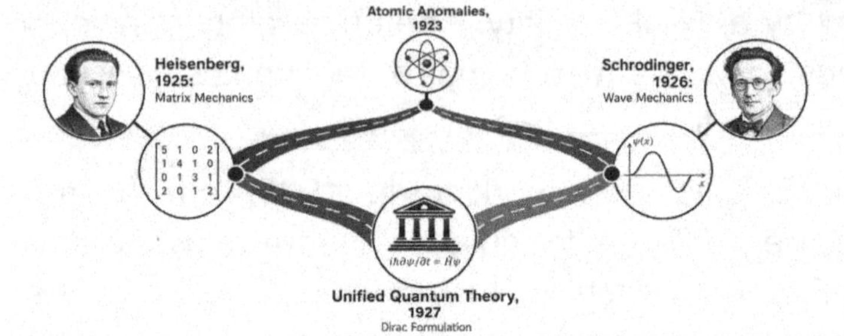


Two physicists discovered the same truth by completely different routes. The roads converge at identical predictions.

## What the Non-Commutativity Means

The technical heart of Heisenberg's discovery is non-commutativity. In ordinary arithmetic, the order of multiplication does not matter. Three times five equals five times three. But for matrices representing position and momentum in quantum mechanics, order matters. Measure position first, then momentum, and you get a different result than measuring momentum first, then position. This is not an artifact of imprecise instruments. It is an algebraic property of the operators representing those quantities.

This non-commutativity is the algebraic signature of incompatibility. When two quantities fail to commute, there is no quantum state in which both are simultaneously sharp. The mathematical proof directly connects to the physical principle: non-commuting operators imply an uncertainty relation between the corresponding physical quantities. This is why the uncertainty principle is not limited to position and momentum or energy and time. Any pair of quantum observables that do not commute will be subject to an uncertainty relation.

The reason position and momentum fail to commute ultimately traces back to the wave nature of particles. In wave mechanics, position is a location in space, while momentum corresponds to the frequency of oscillation of the wave. Sharpen the location, and you must broaden the frequency. Sharpen the frequency, and the wave must extend across all of space. These two operations are genuinely incompatible. The mathematics of Fourier analysis, the same mathematics used to decompose sound waves into their component frequencies, dictates the trade-off. Heisenberg did not impose a limit on nature from the outside. He discovered a limit that was already encoded in the wave structure of everything.

## Common Misconceptions

### The Observer Effect Is Not the Uncertainty Principle

Perhaps the most persistent confusion about Heisenberg's principle is the belief that it is about the observer effect: the idea that the act of measuring something necessarily disturbs it. This confusion arises partly from the gamma-ray microscope thought experiment and partly from loose popular writing. The observer effect is real in quantum mechanics, but it is a different phenomenon. The uncertainty principle would exist even if we could observe particles without disturbing them at all, because it arises from the wave structure of quantum states, not from the mechanics of measurement interactions.

Imagine a hypothetical world in which you could observe an electron's position without kicking it at all. The uncertainty principle would still apply. The electron's wave function would still be spread over a range of momenta. Measuring its position would still leave its momentum undetermined, not because measurement disturbed it, but because the wave function representing it cannot simultaneously be a sharp position spike and a single definite momentum. The disturbance interpretation is a useful heuristic for the gamma-ray microscope, but it is not the foundation of the principle.

Physicists draw this distinction carefully. The Heisenberg uncertainty principle concerns the quantum state itself: the product of the position and momentum uncertainties for any quantum state. A separate relation, sometimes called the Kennard or Robertson uncertainty relation, applies to any pair of non-commuting observables. These are properties of quantum states, not of measurement procedures. A related but distinct concept, the measurement-disturbance relation, concerns how much measuring one quantity actually disturbs the other in real interactions. These three things are related, but not the same, and conflating them leads to the popular but incorrect narrative that quantum mechanics is just about clumsy experimenters.

## Consciousness Does Not Cause Collapse

A second misconception extends the observer effect in a direction that neither Heisenberg nor Bohr intended. Because quantum mechanics uses the word 'observer' and measurement seems to play a special role in the theory, some popular accounts have concluded that conscious observation collapses the wave function and that human consciousness is therefore fundamental to quantum reality. This interpretation has no support in physics. An 'observer' in quantum mechanics is any physical system that interacts with and records information about a quantum state. A photographic plate, a CCD sensor, a cloud chamber, or a molecule in the atmosphere all serve as observers. No consciousness is required.

The uncertainty principle, in particular, has nothing to do with consciousness. It applies to electrons in the interior of stars, to processes that occurred before any biological life existed, to quantum fluctuations in the first moments of the universe. The principle is a structural feature of quantum mechanics, not a statement about the role of minds in physics. This clarification does not make the principle less profound. If anything, the fact that reality is genuinely indeterminate at the quantum level is more unsettling than the idea that human attention is required to make things definite.

Diagram 4.11 -- Common Misconceptions About the Uncertainty Principle

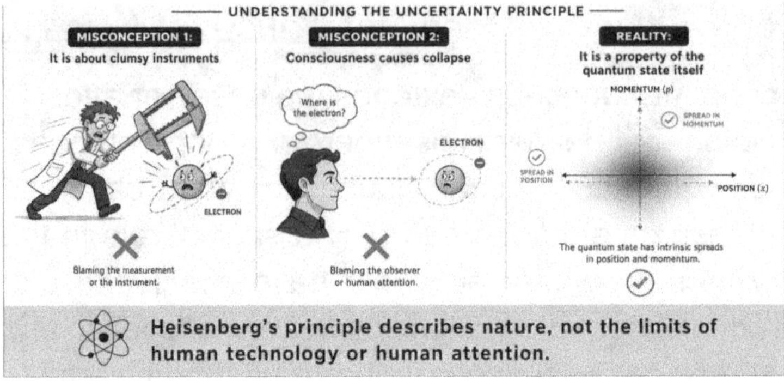


## The Philosophical Weight

### Indeterminism as a Feature, Not a Bug

Before quantum mechanics, the dominant picture of physics was deterministic. Given the exact positions and momenta of every particle in the universe at a single moment, classical mechanics, in principle, claimed to predict the entire future and reconstruct the entire past. Laplace imagined a demon with this complete knowledge, effortlessly calculating the trajectory of every atom from the Big Bang to the end of time. The uncertainty principle permanently closes the door on this vision. It is not that the demon lacks computing power or measurement precision. The demon cannot have complete initial conditions because such conditions do not exist. Position and momentum are not both defined with infinite precision for any particle. There is no complete specification of the quantum state that would allow perfect classical prediction, because the classical variables are not simultaneously real.

This is not merely a statement about electrons. It percolates upward through chemistry, through the firing of neurons, through the weather, into every complex system in the universe. Quantum indeterminism is not merely a curiosity confined to physics laboratories. It is woven into the fabric of causation at the most fundamental level. Whether this indeterminism has consequences for human free will, for the meaning of probability, or for the interpretation of scientific law remains a subject of debate among physicists and philosophers. The physics itself does not answer them, but it makes them unavoidable.

## What Remains Fixed

The uncertainty principle is not a statement of total ignorance. It is a precise, quantitative trade-off. It specifies exactly how much knowledge can coexist about complementary variables. In that sense, it defines rather than destroys the limits of what can be known. A quantum state is fully characterized by its wave function. Everything that can be known about the system is contained in that function. The uncertainty principle tells you that this complete information does not include simultaneously sharp values of both position and momentum. It does not say that quantum mechanics is vague or indeterminate in all respects. The energy levels of hydrogen are known to an extraordinary degree of precision. The behavior of electrons in semiconductors is predicted and engineered with breathtaking accuracy. Quantum mechanics is not imprecise. It is precisely a theory of which things are definite and which are not.

This is the mature understanding that Heisenberg's work produced: not confusion, but a precise map of which questions have sharp answers and which do not. The fog of certainty that this chapter's title invokes is not random ignorance. It is a structured, quantifiable indeterminacy, with a firm mathematical outline and a rich set of physical consequences. The investigator who grasps this is not confused about quantum mechanics. They have understood one of its deepest features.

## Takeaway

Heisenberg's uncertainty principle is not about clumsy instruments, insufficient technology, or the disturbance caused by observation. It is a statement about the quantum state itself. Position and momentum cannot both be sharply defined simultaneously because quantum particles are waves, and a localized wave requires a superposition of many wavelengths, each corresponding to a different momentum. The more precisely you fix one, the more broadly spread the other must become. This trade-off is encoded in the algebraic structure of quantum mechanics: position and momentum are represented by noncommuting operators, and noncommuting operators cannot both be simultaneously sharp.

The same logic applies to energy and time. Short-lived quantum states have large energy spreads. This makes the vacuum itself unstable to quantum fluctuations: virtual particle pairs briefly borrow energy from the vacuum, producing measurable effects, including the Casimir force and the Lamb shift. The uncertainty principle also stabilizes atoms: tight confinement near the nucleus would require enormous kinetic energy, so the ground state is a balance point rather than a collapse. Without this principle, matter itself would be impossible.

Heisenberg discovered the principle through matrix mechanics, and Schrodinger found the same physics through wave mechanics. Both routes were confirmed by experiment and shown to be mathematically equivalent. The uncertainty is ontological, not epistemic: the electron before measurement does not have a hidden definite position that we are failing to see. The quantum state is genuinely spread. This is not a limitation of human knowledge. It is the structure of reality at the smallest scales, precise in its imprecision, and foundational to the world you inhabit.

# 5 Both at Once: The Radical Logic of Superposition

Picture a coin spinning in the air. You might say it is neither heads nor tails right now, and in a sense, you would be correct. But the moment it lands, it will be one or the other. Classical physics says the exact angle, force, and wobble of your thumb always determine the outcome. If you knew every variable, you could predict it. The spinning coin is not genuinely undecided. It looks that way only because you lack information.

Now imagine a quantum particle in a superposition of two states. The temptation is to say the same thing: the particle is in one state or the other, and we do not know which. That temptation leads directly to an error. Quantum mechanics says something far stranger. The particle is not hiding its state from you. It genuinely has no definite state. It is both at once, not because of your ignorance, but because nature has not yet settled the question. The investigator's job in this chapter is to understand what that claim actually means, to see the evidence behind it, and to recognize why it is one of the most consequential facts in science.

## Why It Matters

Superposition is not a quirk confined to physics laboratories. It is the reason quantum computers can outperform classical machines on certain problems and part of the explanation for why plants harvest sunlight with near-perfect efficiency. It is the concept

Erwin Schrodinger tried to show was absurd, and which turned out to be precisely correct. Without understanding superposition, the rest of quantum mechanics is a collection of formulas without a story.

## The Classical Intuition and Where It Breaks

### Heads, Tails, and Hidden Information

The spinning coin is the most natural first image for superposition, and the most important one to get exactly right before you discard it. A spinning coin looks indeterminate. You genuinely cannot tell which face will land up while it is still in the air. But ask a physicist with precise instruments to measure the coin's position, velocity, and angular momentum at any moment during the spin, and she could, in principle, calculate the outcome. The information was always there. The coin's condition was always definite. You did not have access to it.

This is the classical picture: apparent uncertainty is always a cover for hidden information. The world is definitely beneath the surface. Your ignorance does not change the underlying reality. The coin is heads or tails; you merely do not know which.

Quantum mechanics breaks this picture cleanly. If a quantum particle were simply in one definite state that you did not know, it would behave differently than if it were genuinely in both states at once. When you test the difference, the particle behaves as though it is genuinely in both states. The hidden-information explanation fails. The particle is not a coin that has already landed. It is something more radical.

Diagram 5.1 - Spinning Coin vs. Quantum Particle: Two Kinds of Uncertainty

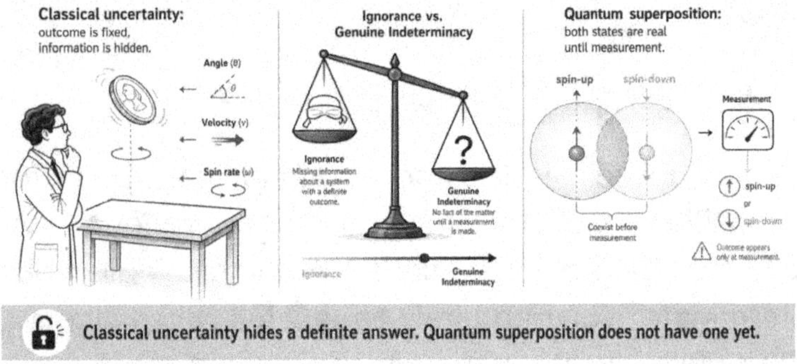


A split panel. On the left, a coin spinning in mid-air above a table, with a physicist figure watching it. A label reads 'Classical uncertainty: outcome is fixed, information is hidden.' Small arrows indicate knowable physical variables such as angle, velocity, and spin rate. On the right, a quantum particle is depicted as two overlapping ghost images, one labeled spin-up and one labeled spin-down, coexisting simultaneously. A label reads 'Quantum superposition: both states are real until measurement.' A connecting scale in the center weighs ignorance against genuine indeterminacy, tipping toward genuine indeterminacy. Style: blue palette, white background, both icons and human figures, flat vector, 16:9 landscape, caption bar reading 'Classical uncertainty hides a definite answer. Quantum superposition does not have one yet.'

The analogy fails for a specific reason. The failure is not just philosophical but experimental. When quantum particles are in superposition, they interfere in ways that can only be explained if both states were genuinely present simultaneously. A coin with a secret head cannot interfere with a coin with a secret tail. That kind of interference does not exist in classical physics. In quantum mechanics, it is routine.

## What Physicists Mean by a State

The word 'state' in physics has a precise meaning: a complete description of a physical system at a moment in time. In classical mechanics, the state of a ball is its position and velocity. Those two numbers tell you everything you need to predict what the ball will do next. The state is singular and definite.

In quantum mechanics, the state of a particle is described by a wave function, which is examined more closely in Chapter 6. A quantum state can be a combination of multiple classical states. An electron can be in a state that combines spin-up and spin-down. A photon can be in a state that combines two paths through an optical instrument. These are not approximations. They are exact descriptions of what the quantum system is doing.

The name for this kind of combination is superposition. The word comes from the Latin for 'placed above,' and in quantum mechanics, it describes states that are layered on top of each other, coexisting until measurement forces the system into one of them. The critical point is that superposition is not a statement about what you know. It is a statement about the particle.

## Electron Spin: The Clearest Window Into Superposition

### What Spin Actually Is

The most straightforward physical example of quantum superposition is the electron's spin. The word 'spin' is borrowed from everyday life but has a quite different meaning in quantum mechanics. An electron does not literally rotate the way a top rotates: no charged object that small could spin fast enough to produce the magnetic properties electrons exhibit without violating known physical laws. Spin is an intrinsic quantum property with no classical equivalent.

What spin does is give the electron a tiny magnetic moment, meaning the electron behaves like an incredibly small magnet. When you probe an electron in an external magnetic field, you find its spin can point in exactly two orientations relative to the field. Physicists call these spin-up and spin-down. There is no continuous range, unlike a classical spinning object. The electron is quantized: spin-up or spin-down when measured. Before measurement, it can be in a superposition of both.

Diagram 5.2 - The Two Faces of Electron Spin

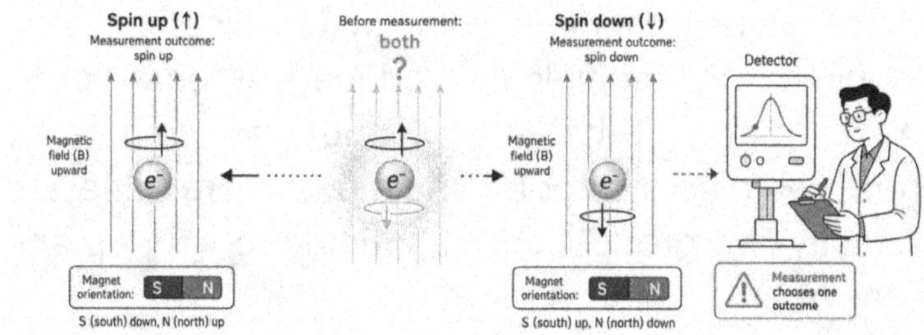


An electron in superposition carries both spin orientations until a measurement chooses one.

A central diagram showing an electron depicted as a small glowing sphere. On the left branch, a magnetic field represented by parallel vertical lines and an upward-pointing arrow indicates spin-up, with a label indicating the orientation of a small bar magnet. On the right branch, the same magnetic field, with an arrow pointing down, indicates spin-down. In the center, between the two branches, a shimmering region shows both arrows overlapping in a single superposition state, with a question mark label reading 'Before measurement: both.' A stylized scientist figure stands beside a detector device, holding a clipboard. Style: blue palette, white background, both icons and human figures, flat vector, 16:9 landscape, caption bar reading 'An electron in superposition carries both spin orientations until a measurement chooses one.'

## The Stern-Gerlach Experiment

In 1922, Otto Stern and Walther Gerlach demonstrated quantum spin in a way no one had anticipated. They fired a beam of silver atoms through an inhomogeneous magnetic field. In classical physics, you would expect the field to produce a smeared band of impact marks. What they found was completely different: two discrete spots. The atoms were deflected into exactly two groups, corresponding to spin-up and spin-down.

The result showed that spin is not a continuous classical quantity. It is quantized. But the deeper implication takes a moment to appreciate. When a silver atom enters the magnetic field in a superposition, the field causes it to register in one of the two spots. Interacting with the field resolves the superposition. This is the first glimpse of what physicists call measurement: the magnetic field does not merely record the spin. It participates in determining the outcome.

Diagram 5.3 - The Stern-Gerlach Apparatus

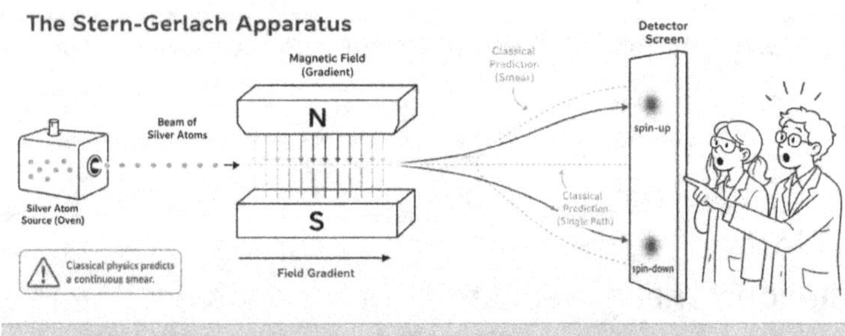


Instead of a smear, Stern and Gerlach found two spots: the first direct evidence of quantized spin.

A horizontal schematic showing a beam of silver atoms traveling from left to right through a pair of magnetic poles, one labeled N and one labeled S, with arrows indicating the field gradient. On the right side, a detector screen shows two discrete spots, one at the top labeled spin-up and one at the bottom labeled spin-down, with a dotted arc showing what a classical smear would look like. A ghostly path in the center represents the classical prediction that did not occur. Two scientist figures stand at the detector, pointing at the two spots with expressions of surprise. Style: blue palette, white background, both icons and human figures, flat vector, 16:9 landscape, caption bar reading 'Instead of a smear, Stern and Gerlach found two spots: the first direct evidence of quantized spin.'

The Stern-Gerlach experiment set the template for how physicists think about quantum measurement. The apparatus does not passively read off a pre-existing value. It actively engages with the system's quantum state. An electron in a superposition of spin-up and spin-down will, when it passes through a Stern-Gerlach magnet, land in one of the two spots with definite probabilities. But before the interaction, neither outcome was guaranteed. Both were equally real, coexisting in the superposition.

## Spin Superposition Is Not Probabilistic Ignorance

The distinction between genuine superposition and probabilistic ignorance is philosophically important, but it is also testable. Take a large number of electrons, each prepared in a superposition of spin-up and spin-down in the vertical direction. Measure their vertical spin. Half will be up, half will be down. So far, this looks like ignorance: maybe they were always up or down, and you did not know which.

Now take another set of electrons in the same superposition state. Instead of measuring vertical spin, rotate the apparatus ninety degrees and measure horizontal spin. Again, you get half up and half down. But now comes the critical test. Take electrons that registered as vertical-up (they are now in a definite vertical state) and measure them in the horizontal direction. Again, half up and half down. Now take those horizontal results and measure them again vertically. What do you get? Half up and half down again.

If spin had been a hidden definite quantity all along, the second vertical measurement would have given the same answer as the first. It does not. Measuring the horizontal spin actually disturbs the vertical state. This sequence is only possible if the electron was genuinely in a superposition of both vertical spin states after the horizontal measurement, because the horizontal measurement put it there. The state is not discovered by measurement. It is, in a very real sense, created by it.

Diagram 5.4 - Sequential Spin Measurements: Proof Against Hidden Variables

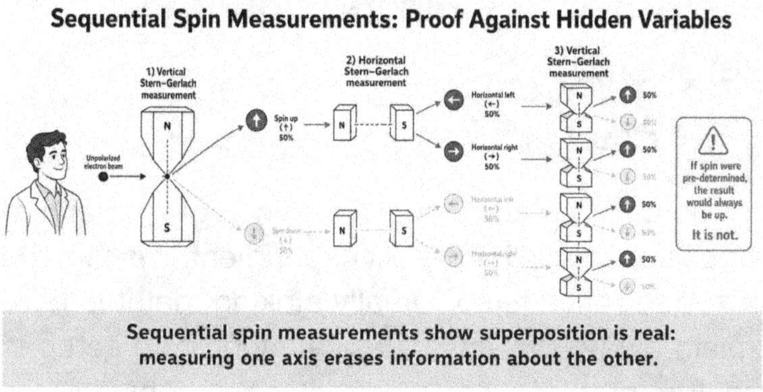


A flowchart of spin measurements arranged left to right. First panel: electron enters a vertical Stern-Gerlach device and splits into spin-up (selected, shown in solid blue) and spin-down (discarded, shown faded). Second panel: the spin-up electron enters a horizontal Stern-Gerlach device, randomly splitting into horizontal-left and horizontal-right. Third panel: the horizontally left electron enters another vertical Stern-Gerlach device and splits 50/50 into spin-up and spin-down. A highlighted annotation reads 'If spin were pre-determined, the result would always be up. It is not.' Style: blue palette, white background, both icons and human figures, flat vector, 16:9 landscape, caption bar reading 'Sequential spin measurements show superposition is real: measuring one axis erases information about the other.'

## Interference: The Fingerprint of Superposition

### The Double-Slit Experiment

The most famous confirmation of quantum superposition is the double-slit experiment. It is famous because the result is so visually striking that it is hard to explain away with any classical story. The setup is simple. Fire particles, say electrons or photons, toward a barrier that has two narrow slits cut into it. Behind the barrier is a detector screen that records the location of each particle.

If you close one slit and fire particles through the other, you get a single band behind the open slit. Classical logic says that with both slits open, you should get two bands. But that is not what you get. You get an interference pattern: alternating bright and dark bands across the screen, like ripples from two stones that overlap and reinforce or cancel each other. The pattern is the unmistakable signature of waves.

Here is the harder part. You get this interference pattern even if you fire the electrons one at a time, waiting long enough between each firing that there is never more than one electron in the apparatus at any moment. Each electron seems to pass through both slits simultaneously, interfering with itself, and to land at a position consistent with the overall wave pattern. Over many electrons, the interference pattern builds up, dot by dot.

Diagram 5.5 - The Double-Slit Experiment: Particles That Interfere With Themselves

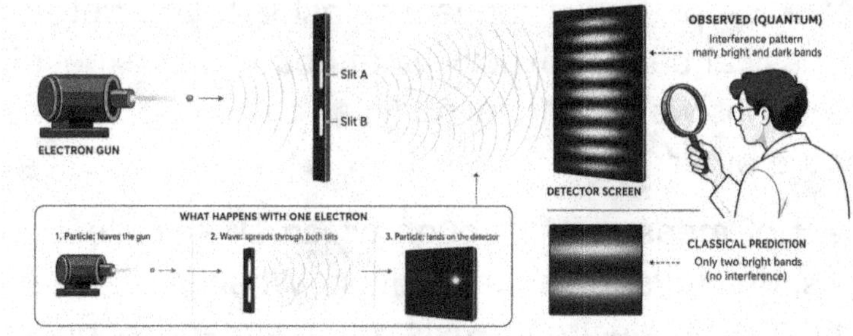


Each electron interferes with itself, building an interference pattern one impact at a time.

A central schematic showing an electron gun on the far left. In the center, a barrier with two narrow vertical slits labeled Slit A and Slit B. On the right, a detector screen showing a classic multi-band interference pattern with alternating bright blue and dark regions. Below the detector, a separate panel shows the classical prediction: only two bright bands. A magnified inset shows a single electron leaving the gun as a particle, spreading through the barrier region as a wave, and landing at the detector as a point. A scientist figures out the interference pattern with a magnifying glass. Style: blue palette, white background, both icons and human figures, flat vector, 16:9 landscape, caption bar reading 'Each electron interferes with itself, building an interference pattern one impact at a time.'

## What Happens When You Watch

The double-slit experiment has a second act that is even more unsettling. Suppose you place detectors at the slits so that you can tell which slit each electron actually passes through. What happens? The interference pattern disappears. Instead, you get the two classical bands, exactly what you would expect if the electrons were ordinary particles going through one slit or the other.

The act of measuring, through which slit the electron passes, destroys the superposition. When you are not watching, the electron exists in a superposition of having-gone-through-slit-A and having-gone-through-slit-B. This superposition allows interference. When you force the electron to register at one slit or the other, it can no longer interfere with itself. The interference pattern is the fingerprint of genuine superposition. Its disappearance when you measure the path confirms that the superposition was real and that measurement disrupts it.

This result is not a failure of the detector or a disturbance from the detector's photons, though that was the initial classical hope. Even when the detectors are made gentle enough to disturb the electron barely, the interference pattern vanishes as long as any which-path information is recorded anywhere in the universe. The universe keeps track. Superposition holds only when no information has been recorded about which alternative was chosen.

Diagram 5.6 - Which-Path Detection Destroys Interference

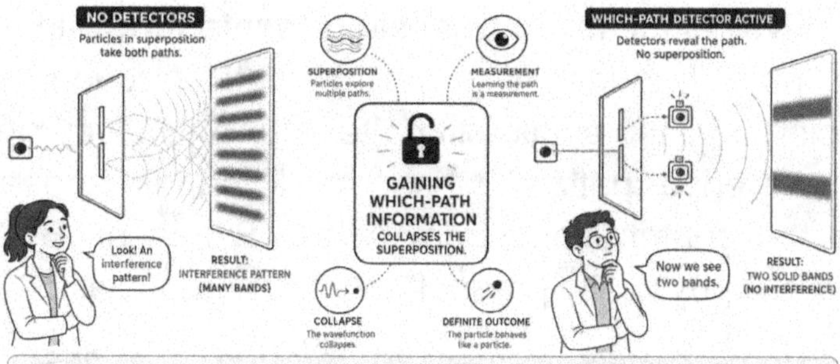


The moment you know which slit the particle took, the interference vanishes.

A two-panel comparison. Left panel: double-slit apparatus with no detectors at slits, and the resulting interference pattern on the screen showing many alternating bands. Right panel: identical apparatus, but now small detector icons are placed at each slit, labeled 'which-path detector active.' The screen on the right shows only two solid bands, the classical particle pattern. Between the panels, a central annotation reads 'Gaining which-path information collapses the superposition.' A pair of scientist figures, one on each side, compare the two screens. Style: blue palette, white background, both icons and human figures, flat vector, 16:9 landscape, caption bar reading 'The moment you know which slit the particle took, the interference vanishes.'

## Interference as Physical Proof

The interference pattern is not a curiosity. It is rigorous evidence that the particle was in a genuine superposition, not a hidden definite state. If the electron had secretly gone through slit A, there would be no physical influence from slit B to interfere with. A secret path through A is just a path through A, producing only the single-slit pattern for A. But you get an interference pattern whose dark bands appear where the contributions from A and B cancel. Both paths were physically present and physically real.

This argument was formalized through a series of increasingly precise experiments over the twentieth century, culminating in experiments in which single photons were sent through kilometers-long interferometers. The superposition persisted across the entire path length. The two states were genuinely coexistent, not merely unknown, not merely unobserved.

## Schrodinger's Cat: The Argument, Not the Punchline

### Erwin Schrodinger's Original Intention

In 1935, Erwin Schrödinger proposed his famous thought experiment not to demonstrate that quantum mechanics yields strange results, but to argue that, taken literally at the macroscopic scale, it leads to an absurdity. He was making a reductio ad absurdum: if your theory implies something obviously false, then something is wrong with your theory.

Schrodinger was deeply troubled by what quantum mechanics seemed to say about measurement. The theory allowed particles to exist in superpositions but said those superpositions collapse into a single outcome upon measurement. Schrodinger wanted to know exactly when and how this collapse happened. He was not satisfied with the prevailing answer, which was essentially: it happens when a measurement occurs. He found that vague and unsatisfying. His thought experiment was designed to make the vagueness visible by scaling the quantum superposition up to a macroscopic level that no one could dismiss.

Diagram 5.7 - Schrodinger's Cat: The Setup and the Dilemma

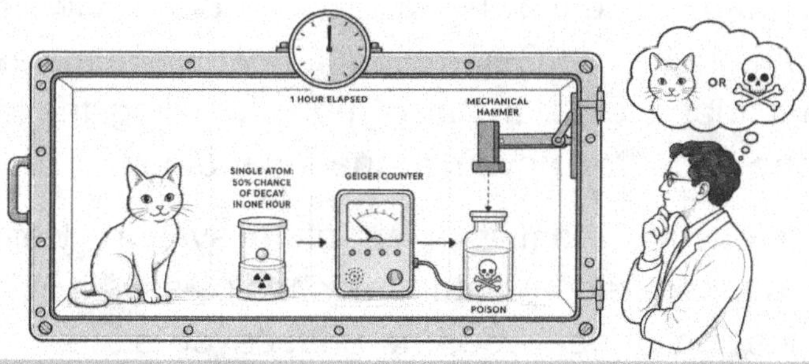


Schrodinger asked: is the cat alive and dead simultaneously, or is quantum superposition limited to the microscopic world?

A cross-sectional diagram of a sealed steel box. Inside the box on the left is a small, stylized cat drawn in a friendly, non-gory way. In the center is a small radioactive sample with a label reading 'single atom: 50% chance of decay in one hour.' Connected to the sample by an arrow is a Geiger counter, which connects by another arrow to a mechanical hammer, which sits above a glass vial labeled 'poison.' At the top of the box, a clock shows that one hour has elapsed. Outside the box, a scientist figure stands with a question-mark thought bubble containing both a live cat and a skull-and-crossbones symbol. Style: blue palette, white background, both icons and human figures, flat vector, 16:9 landscape, caption bar reading 'Schrodinger asked: is the cat alive and dead simultaneously, or is quantum superposition limited to the microscopic world?'

## The Chain of Entanglement

The setup goes like this. Place a cat in a sealed steel box. Inside is a tiny amount of radioactive material with a 50% chance of decaying within one hour. If the atom decays, it triggers a Geiger counter, which releases a hammer that breaks a vial of poison, killing the cat. After one hour, you have not opened the box.

The radioactive atom is a quantum system. Before decay is measured, quantum mechanics says it is in a superposition of decayed and undecayed states. The atom is quantum-mechanically linked, through the chain of Geiger counter, hammer, vial, and poison, to the cat. If the atom is in a superposition and the chain faithfully transmits the quantum state, then the cat must also be in a superposition of alive and dead. But that is plainly ridiculous. A cat cannot be both alive and dead.

Schrodinger was not saying the cat is both alive and dead. He was saying: quantum mechanics, as then formulated, seems to require you to say that, and since that is obviously wrong, the formulation must be incomplete. The thought experiment was an argument for the need for a more complete theory, not a declaration of quantum weirdness.

## What Schrodinger's Cat Actually Teaches

The thought experiment has been misread in popular culture, where it is often presented as proof that the cat really is both alive and dead until you look. Schrodinger himself would have been appalled at that reading. His point was precisely the opposite: macroscopic objects like cats cannot be in superpositions of grossly different states, so there must be a boundary somewhere between the quantum world where superposition operates and the classical world where definite states prevail.

Quantum mechanics, as developed through the twentieth century, vindicates Schrödinger's concern, though not in the way he imagined. The process by which quantum superpositions dissolve when they interact with large environments is called decoherence. The cat is not in a superposition because the decaying atom is not isolated: it interacts with the environment inside the box, and those interactions destroy the superposition far faster than any human measurement could observe. The cat's fate is settled long before you open the box. But at the microscopic scale, where systems can be isolated, superposition is real and persists. Schrodinger's cat remains useful because it forces the right questions: What counts as a measurement? Where does the quantum world end and the classical world begin?

Diagram 5.8 - Decoherence: How the Environment Resolves the Superposition

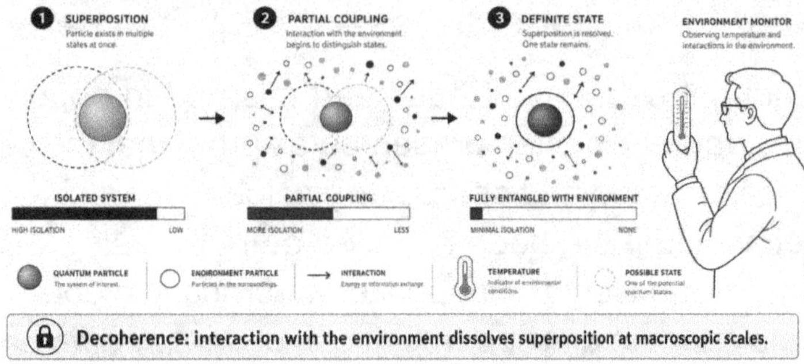


A sequence of three panels from left to right. Panel 1: a single quantum particle shown as a superposition of two states, depicted as two ghost outlines overlapping. Panel 2: The particle is surrounded by a cloud of environmental particles representing air molecules, photons, and heat. The superposition outlines begin to separate and dim. Panel 3: The environmental entanglement has resolved the system into a single definite state. Each panel shows the level of quantum isolation, with bars decreasing in length labeled 'isolated system,' 'partial coupling,' and 'fully entangled with environment.' A scientist figures with a thermometer monitors the environment. Style: blue palette, white background, both icons and human figures, flat vector, 16:9 landscape, caption bar reading 'Decoherence: interaction with the environment dissolves superposition at macroscopic scales.'

# The Measurement Problem

## What Happens When You Look

The measurement problem is the deepest unresolved question in the interpretation of quantum mechanics. Here is the problem stated plainly. Before measurement, quantum mechanics describes a particle as being in a superposition of states. The mathematical formalism evolves this superposition smoothly and continuously according to a precise equation. But after measurement, we find the particle in one definite state. The two descriptions are in tension.

The classical Copenhagen interpretation addressed this by saying that the wave function collapses upon measurement. Do not ask what happened during the collapse or why; only accept that it happened. When measurement occurs, the superposition ends, and one outcome is selected according to the probabilities. This answer is pragmatically useful but philosophically unsatisfying because it introduces 'measurement' as a primitive term without defining it precisely.

Diagram 5.9 - The Measurement Problem: Before and After

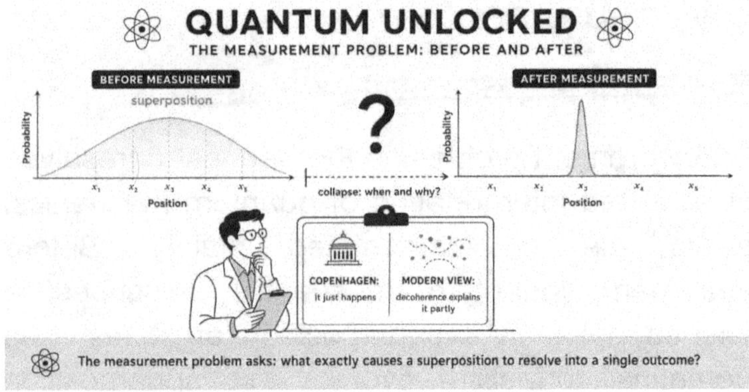


A diptych panel. On the left, labeled 'Before measurement,' a particle is shown as a smooth wave distribution spreading across multiple positions, all equally real, under the heading 'superposition.' On the right, labeled 'After measurement,' the same particle collapses into a single sharp peak at a single location. Between the panels, a large question mark represents the boundary, with a stylized timeline gap labeled 'collapse: when and why?' At the bottom, a scientist figure holds a clipboard with two columns: 'Copenhagen: it just happens' and 'Modern view: decoherence explains it partly.' Style: blue palette, white background, both icons and human figures, flat vector, 16:9 landscape, caption bar reading 'The measurement problem asks: what exactly causes a superposition to resolve into a single outcome?'

## Three Interpretations in Brief

Physicists have proposed several interpretations of quantum mechanics that handle the measurement problem differently. None has been experimentally confirmed over the others, because they all make identical predictions for observable outcomes. They differ only in what they say is really happening beneath the surface.

The Copenhagen interpretation says: stop asking what is really happening and calculate. The wave function is a tool for probability, not a map of reality. Measurement collapses it, and the collapse is the end of the story. The many-worlds interpretation, proposed by Hugh Everett in 1957, says there is no collapse: the universe literally branches. In one branch, the electron is spin-up; in another, it is spin-down, and we see only one outcome because we are located in one branch.

The pilot-wave interpretation, associated with David Bohm, says the particle always has a definite position, guided by a real wave that explores all paths simultaneously. All three interpretations make the same predictions and are taken seriously by some physicists. The measurement problem remains genuinely open.

## Decoherence: The Modern Partial Answer

The most significant advance in understanding the measurement problem since the 1970s is the theory of decoherence. Decoherence does not solve the measurement problem, but it substantially reduces it. The key insight is that real quantum systems are never perfectly isolated. They interact with their environment: air molecules, photons, heat, vibrations. Each interaction creates an entanglement between the quantum system and its environment. Once a particle is entangled with trillions of environmental particles, its quantum state becomes indistinguishable from a classical mixture of definite states.

The timescale for decoherence depends on the system size and its coupling to the environment. For a dust particle in ordinary air, decoherence takes roughly ten to the minus thirty-one seconds. For a single atom in an ultrahigh-vacuum environment, decoherence can be suppressed long enough to run experiments. Decoherence explains why the macroscopic world appears classical. It does not fully explain why a specific outcome is selected; that remains the hard core of the measurement problem. But it dissolves the mystery of why we never see cats in superpositions.

## Macroscopic vs. Quantum Superposition

### Why Cats Don't Sit in Both States

The distinction between macroscopic and quantum superposition is fundamentally a distinction in isolation. A single electron in a well-controlled laboratory can be maintained in a superposition of spin states for microseconds or longer, provided it is carefully cooled and shielded. A cat cannot. The difference is not that the cat is made of special non-quantum material. Cats are made of atoms, and atoms obey quantum mechanics. The difference is scale and coupling to the environment.

A cat contains roughly ten to the twenty-seventh atoms, each interacting with all its neighbors via thermal vibrations, electromagnetic forces, and chemical bonds. The quantum state of any one atom is immediately entangled with all the others, and through them with the surrounding air, the walls of the box, and the cosmic microwave background. This web of entanglement spreads so quickly that any coherent superposition is dissipated in a time far too short to detect. Quantum mechanics applies to cats, to planets, to galaxies. But macroscopic superpositions of grossly distinct states are suppressed by decoherence so completely that they are physically indistinguishable from definite classical states.

Diagram 5.10 - Scale and Decoherence Time: Why Quantum Becomes Classical

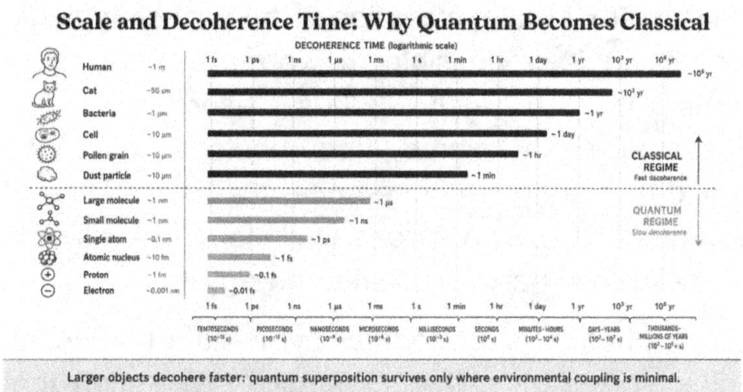


A vertical chart showing physical systems arranged from bottom to top in increasing size: electron, proton, atomic nucleus, single atom, small molecule, large molecule, dust particle, pollen grain, cell, bacteria, cat, human. For each system, a horizontal bar extending to

the right indicates the decoherence time on a logarithmic scale, ranging from femtoseconds to millions of years. Single electrons and atoms show long coherence times. Large molecules show much shorter times. A dust particle shows essentially zero coherence time. A dividing line somewhere around the single-atom level marks the quantum-to-classical boundary. Style: blue palette, white background, both icons and human figures, flat vector, 16:9 landscape, caption bar reading 'Larger objects decohere faster: quantum superposition survives only where environmental coupling is minimal.'

## Quantum Superposition at the Biological Scale

There is a remarkable exception to the expectation that superposition is impossible in warm, wet biological systems. In the early 2000s, experiments on photosynthesis in plants and bacteria produced evidence that quantum coherence, the persistence of quantum superposition across multiple states, plays a functional role in harvesting solar energy.

When a photon is absorbed by a light-harvesting complex in a plant cell, it creates an excitation that must travel from antenna molecules to the reaction center. Classical physics predicts random wandering through the molecular network. Instead, experiments show that the excitation explores multiple paths simultaneously via quantum interference, finding the most efficient route far more reliably than any classical random walk. The coherence persists for hundreds of femtoseconds, long enough to matter.

The biological relevance of this quantum coherence is still debated. Some researchers argue that the coherence is incidental, a side effect of molecular structure rather than a designed feature. Others argue that it is a genuine quantum optimization problem. Nature has, at minimum, found ways to exploit coherence in warm, noisy biological environments for functional purposes, suggesting that quantum effects may be more widespread in biology than anyone expected.

## Qubits: Superposition as a Resource

### Classical Bits and Their Limitations

A classical computer performs calculations using bits. A bit is the simplest unit of information: it is either zero or one. Every classical computation is a sequence of operations on these binary values. The limitation of classical bits is that each bit is always in exactly one state. To check multiple possibilities simultaneously, a classical computer must check them sequentially. Some problems require checking an astronomical number of possibilities, and even the fastest classical computers take impractical amounts of time.

A qubit, the quantum analog of a bit, can be in a superposition of zero and one simultaneously. This is not a metaphor. The qubit genuinely exists in both states at once until measured. Because a quantum computer can maintain the superposition across many qubits simultaneously, it can represent and process an exponential number of states at the same time.

## What Qubits Can and Cannot Do

The power of quantum computation stems from the fact that a superposition of qubits can be manipulated as a whole, performing operations on all states in parallel. This is not the same as a classical computer running many calculations on separate processors. It is more subtle: a single quantum system whose state encodes all possibilities at once, and whose evolution under quantum gates performs a simultaneous computation on all of them.

This power is not unlimited. At the end of a quantum computation, you must measure the qubit register, and measurement collapses the superposition to a single outcome. If you designed your computation cleverly, the interference of the quantum states during the computation amplified the probability of the correct answer and suppressed the probability of incorrect answers. That is what quantum algorithms are: careful orchestrations of interference to make the right answer come out with high probability.

Two examples clarify the scope of quantum advantage. Shor's algorithm, developed in 1994, can factor large numbers exponentially faster than any known classical algorithm, which matters because much internet encryption relies on the difficulty of factoring. Grover's algorithm offers a quadratic speedup for searching unsorted databases. These are real, proven advantages for specific problems. For most everyday computing tasks, a quantum computer offers no advantage and is more fragile and error-prone than a classical machine. Superposition is a resource for specific computations, not a universal accelerator.

Diagram 5.11 - Qubit Superposition and the Bloch Sphere

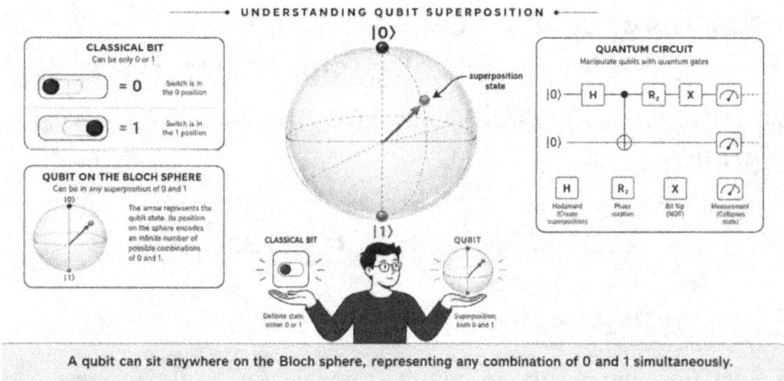


A three-dimensional sphere, the Bloch sphere, is in the center. The north pole is labeled '|0>' in blue and the south pole '|1>' in a slightly darker blue. An arrow from the center points to a position on the surface between the two poles, labeled 'superposition state.' Surrounding the sphere are three inset panels: on the left, a classical bit icon showing a switch in either position 0 or 1; in the center, the qubit arrow on the sphere showing the full range of states; on the right, a simplified quantum circuit diagram with gate symbols. At the bottom, a stylized computer figure holds both a classical bit and a qubit, contrasting their states. Style: blue palette, white background, both icons and human figures, flat vector, 16:9 landscape, caption bar reading

'A qubit can sit anywhere on the Bloch sphere, representing any combination of 0 and 1 simultaneously.'

Today's quantum computers are noisy and limited, operating with dozens to a few hundred qubits before errors overwhelm the computation. Error-correction schemes can, in principle, protect quantum information from decoherence, but scaling to the millions of logical qubits required for practical computations remains a major engineering challenge. The physics is solid. The engineering is hard.

## Quantum Coherence in Photosynthesis

### Light Harvesting and the Quantum Path

Photosynthesis is the mechanism by which plants, algae, and photosynthetic bacteria convert sunlight into chemical energy. Under ideal conditions, nearly every photon absorbed results in successful charge separation at the reaction center, making it one of the most efficient energy-conversion processes in nature.

The light-harvesting complexes in photosynthetic organisms contain tightly packed networks of chromophores, molecules that absorb light. When a chromophore absorbs a photon, it creates an exciton that must hop through the network to the reaction center before the energy is lost as heat or fluorescence. The mystery was why this process is so efficient, given the many possible paths with different lengths and energy levels.

## Wave-Like Energy Transport

What researchers found when they used ultrafast laser spectroscopy to probe the energy transfer process in real time was oscillating quantum beats, signatures of quantum coherence persisting across multiple chromophores simultaneously. The exciton was not hopping randomly from one chromophore to the next like a classical particle. It was spread across multiple chromophores in a superposition, simultaneously exploring multiple paths through the network and using quantum interference to find the most efficient one.

The measured coherence time was on the order of several hundred femtoseconds, roughly as long as the energy-transfer process itself, suggesting that the coherence was not merely incidental but potentially functional. The debate remains lively. Some later experiments suggested that the observed beats arise from vibrational dynamics rather than electronic coherence alone. Others found evidence for a coupled vibrational-electronic coherence that may be more robust than either alone.

What is clear is that quantum superposition is not a fragile laboratory curiosity. It is a physical resource that nature has been exploiting for billions of years. When systems are sufficiently isolated from disruptive environmental noise, superposition persists and does real work. Understanding how biological systems manage this at room temperature remains one of the most exciting open questions at the intersection of quantum physics and biology.

## Takeaway

Superposition is the most fundamental departure from classical intuition in quantum mechanics. A quantum particle in superposition is not hiding a secret, definite value. It genuinely exists in multiple states at once, and that coexistence has measurable, verifiable physical consequences in the form of interference patterns that cannot be explained any other way.

Electron spin is the clearest example: an electron prepared in a superposition of spin-up and spin-down will produce interference effects that collapse the moment any which-path information is recorded. The Stern-Gerlach experiment confirmed that spin is quantized. Sequential spin measurements confirmed that the definite state is created by measurement, not discovered by it.

Schrodinger's cat was not a demonstration of quantum weirdness but an argument against the naive application of superposition to macroscopic objects. Decoherence provides a large part of the answer: large objects coupled to their environments cannot maintain superpositions because quantum information spreads irreversibly outward. The cat's fate is settled before any human looks.

The measurement problem remains genuinely open. No interpretation of quantum mechanics has been experimentally distinguished from the others. They make identical predictions but tell very different stories about what is really happening. Hold your preference with intellectual humility.

Superposition is a physical resource. Quantum computers exploit it to perform computations that would be exponentially difficult for classical machines. Photosynthetic organisms may exploit it to route solar energy with near-perfect efficiency. As quantum technology matures, controlling superposition states and combating decoherence will be the central engineering challenge for the field.

The investigator's lesson from this chapter is precise: when you hear someone say a particle is 'both states at once,' that is not a loose metaphor for uncertainty. It is a literal physical claim, confirmed by interference experiments of extraordinary precision. The particle does not know its state until something makes it choose.

# 6 The Ghost in the Machine: Wave Functions, Probability, and the Shape of the Possible

Imagine you are given a sealed envelope. Inside that envelope is a number. You have not opened it. You do not know whether the number is odd or even, large or small, positive or negative. Now here is the strange question that quantum mechanics forces you to sit with: before you open the envelope, does the number have a definite value at all? Most of us would say yes, obviously. The number is whatever it is. Our ignorance does not change the fact. The envelope is not magic. The number is there, fixed and waiting. Classical physics agrees with you completely.

But in the quantum world, that comfortable assumption breaks down. There is a class of physical quantities, things like the position of an electron or the energy level of an atom, where the system does not have a definite value before you measure it. The system is not hiding something from you. It is genuinely, physically undecided. And the tool that quantum mechanics uses to describe that undecidedness is called the wave function. Understanding the wave function means understanding why the quantum world is not just hard to know but fundamentally different from anything we encounter at the human scale. That is what this chapter is about. Put on your investigator's hat. The evidence is unusual, but it is real.

## Why It Matters

A wave function governs every atom in your body. Every chemical bond that holds your molecules together, every photon emitted by the screen you might be reading this on, every magnetic resonance image taken in a hospital: all of these depend on the wave function behaving exactly as quantum mechanics predicts. When chemists design new drugs or materials scientists engineer new semiconductors, they are, in a very practical sense, working with wave functions. The concept is not an abstraction reserved for theorists. It is the engine of modern science and the hidden logic beneath the technologies that define contemporary life. Before any of that can make sense, though, we need to understand what a wave function actually is.

## From Orbits to Clouds: Why the Old Picture Had to Go

To appreciate what the wave function tells us, it helps to understand what physicists believed before its introduction. In the early twentieth century, the standard picture of the atom was essentially a miniature solar system. The nucleus sat at the center like a sun, and electrons orbited around it in neat circular paths, much as planets circle a star. Niels Bohr formalized this picture in 1913, and for a while it worked surprisingly well for hydrogen, the simplest atom of all.

But the solar system model had deep problems. It could not explain why atoms did not collapse. An orbiting electron, by the rules of classical electromagnetism, should continuously radiate energy and spiral inward into the nucleus in a fraction of a second. Every atom in the universe should self-destruct almost instantly. Clearly, that does not happen. The model also failed to predict the spectral lines of anything more complex than hydrogen. Add a second electron, and the whole neat orbital picture fell apart. Physicists knew by the early 1920s that something fundamental was missing.

The missing piece was not a correction to the solar system model. It was a completely different way of thinking about where electrons are and what "where" even means at the quantum scale. The solution came from an Austrian physicist named Erwin Schrodinger, building on the insight of Louis de Broglie that particles could behave like waves. In 1926, Schrödinger wrote down an equation describing how a quantum system evolves. The solution to his equation was what we now call the wave function.

Diagram 6.1 – From Bohr Orbits to Probability Clouds

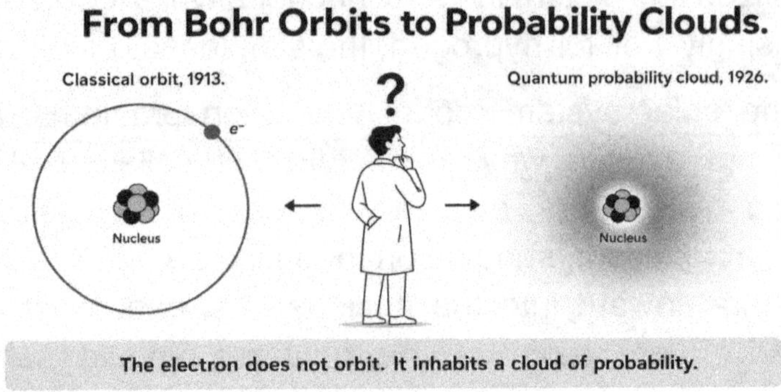


The wave function did not describe where an electron is. It described the full range of possible locations for an electron, each location weighted by its probability. The electron does not have a hidden precise position that we fail to measure. According to quantum mechanics, it genuinely occupies a range of positions simultaneously until a measurement is made. This is a radical departure from anything in classical physics, and it is worth sitting with for a moment before moving on.

## What "Wave" Means Here

The word "wave" in wave function does not refer to a ripple in water or a sound wave traveling through air. Those are waves in a physical medium: actual disturbances moving through space. The quantum wave function is different. It is a mathematical object that encodes information about a particle's possible states. It spreads out through space in a wave-like pattern, but it is not a physical substance you could ever touch or directly detect. What you detect is always a particle at a particular location. The wave function is the ghost of that particle's possibilities, hovering behind the scenes until measurement brings one possibility to the surface.

This distinction matters. Many popular accounts of quantum mechanics describe particles as "really being waves," as if an electron is a smeared-out physical blob undulating through space. That image is misleading. The electron is still a particle when you detect it: a point-like object registering in a specific detector at a specific moment. The wave function describes the probability structure governing where and when that detection will occur. The wave is in the mathematics, not in the stuff.

## Amplitude, Phase, and the Architecture of Possibility

The wave function has two main features that carry physical meaning: amplitude and phase. Both are unfamiliar in everyday experience, but each has a concrete interpretation that makes sense once you have the right analogy.

### Amplitude: How Likely Is It?

The amplitude of a wave function at a given location is a number that tells you, in rough terms, how "present" the particle is at that place. A large amplitude means the particle is more likely to be found there. A small amplitude means it is less likely. The amplitude can be large in one region and small in another, meaning the particle has a high probability of being found in the first region and a low probability in the second. This is the core physical content of the wave function: it maps out the probability landscape of a particle's possible locations.

Here is an analogy. Consider a weather forecast that shows a storm system might make landfall anywhere along a stretch of coastline. The forecast does not say the storm will hit precisely at point X. It gives a probability distribution: higher chance here, lower chance there, near zero at the far ends of the range. The wave function is the forecast applied to a quantum particle. The storm does not have a secret precise landing spot that the meteorologist is merely failing to calculate. And in quantum mechanics, the particle does not have a secret precise location either. The probability is the most complete description available and the only one that exists.

Diagram 6.2 – The Wave Function as a Probability Landscape

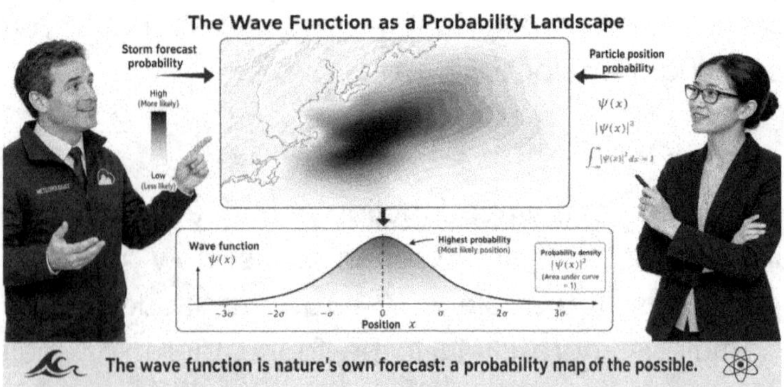

This analogy has a limit, and it is important to note. In weather forecasting, the storm really does have a definite location at every moment. Our uncertainty is due solely to imperfect data and computational limitations. If we had a perfect global sensor network and unlimited computing power, the forecast would sharpen to a single point. In quantum mechanics, that sharpening is not available in principle. The Heisenberg Uncertainty Principle, which you encountered in Chapter 5, is not an engineering problem awaiting a better instrument. The probability spread in the wave function is irreducible. This is where the analogy breaks down, and quantum mechanics parts ways with everything classical.

## Phase: The Hidden Face of the Wave

The second feature of the wave function is phase. If amplitude tells you how likely an outcome is, phase is subtler. It does not directly tell you a probability. Instead, it controls how different parts of the wave function add together or cancel each other out, a phenomenon called interference.

Think of two water waves meeting on a pond. When two crests arrive at the same point simultaneously, they reinforce each other and produce a taller crest. When a crest meets a trough, they cancel each other out, and the water is momentarily still. This is wave interference, and it depends entirely on the relative phase of the two waves. The wave function behaves the same way. Two particle paths can interfere constructively, making that outcome more likely, or destructively, canceling it out and making it impossible.

This is how the double-slit experiment, discussed in Chapter 4, produces its mysterious interference pattern on the detector screen. The wave function of a single particle travels through both slits simultaneously. The two parts of the wave function, one from each slit, arrive at the detector wall with different phases depending on the exact path length. Where they are in phase, they reinforce, and particles accumulate. Where they are out of phase, they cancel, and particles never arrive. The result is stripes of presence and absence: the interference pattern. Phase is what makes the quantum world genuinely wave-like, even though particles are still detected as particles.

Diagram 6.3 – Phase and Interference: How Possibilities Add and Cancel

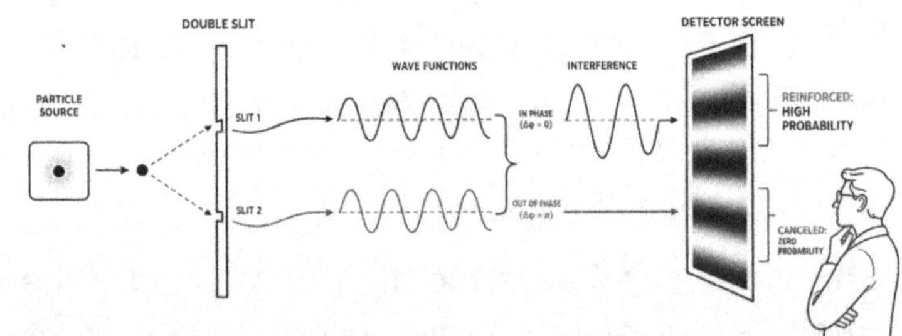


Phase determines which futures reinforce and which cancel.

One important clarification: phase differences are only detectable through interference. If you have a single wave function and no second wave function to compare it to, the absolute phase is unmeasurable. What matters physically is the relative phase between parts of the wave function when they combine. This is one of the more counterintuitive aspects of quantum mechanics: a quantity that influences observable outcomes cannot itself be directly observed. The world is more subtle than our measuring instruments, even in principle.

## Born's Rule: Turning the Wave Into a Probability

At this point, you might be asking: if the wave function is not directly observable, how do we connect it to the actual outcomes of experiments? This is exactly what troubled physicists in the 1920s. Schrödinger had an equation that produced wave functions. But what did the wave function mean for an actual measurement?

The answer came from the German physicist Max Born in 1926, the same year Schrodinger published his equation. Born proposed a conceptually simple rule, even if its physical meaning remains debated to this day. His rule is this: to find the probability of observing a particle at a particular location, take the amplitude of the wave function at that location and square it. The square of the amplitude gives you the probability.

That is Born's rule in its core form: probability is the square of the amplitude. Notice that squaring a number always gives a positive result (or zero), which is exactly what you need for a probability, because probabilities must be between zero and one. The wave function itself can have negative or complex values, which is how phase works. But when you square it to get the probability, the result is always non-negative. Born's rule is the bridge between the abstract mathematics of the wave function and the concrete numbers you read off a detector.

Diagram 6.4 – Born's Rule: From Wave to Probability

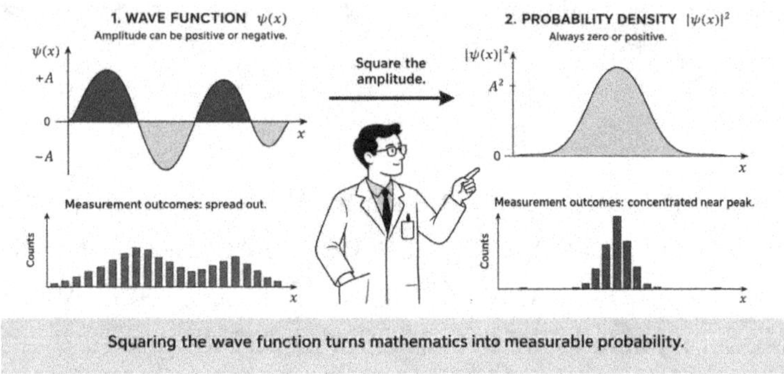


Why squaring? Physically, this is where quantum mechanics genuinely departs from any classical intuition. There is no deeper classical reason why the rule should be squaring rather than cubing or something else. Born's rule is a postulate: a foundational assumption of the theory that is justified solely by its success. Every prediction made using Born's rule has been confirmed by experiment to extraordinary precision. The rule is, as far as we know, exactly correct.

## The Rule in Action: What the Atom Really Looks Like

When you apply Born's rule to Schrodinger's equation for the hydrogen atom, you get something remarkable. The electron in a hydrogen atom does not orbit at a fixed radius like a planet circling the sun. Instead, its wave function is a solution to Schrodinger's equation that spreads out through three-dimensional space in a specific pattern. Squaring that wave function, using Born's rule, gives you a three-dimensional probability distribution: a map showing the relative likelihood of finding the electron at each point in space.

That probability distribution is what chemists call an atomic orbital. The shapes of atomic orbitals are direct predictions of quantum mechanics. They are not artistic approximations or simplified diagrams. They are the actual probability clouds that Born's rule derives from Schrodinger's equation. The familiar s-orbital, which looks like a sphere, means the electron is most likely to be found anywhere on a spherical shell at a particular radius, with equal probability in all directions. The p-orbital, which has a dumbbell shape, means the electron's probability is concentrated in two lobes pointing in a specific direction. These shapes are not metaphors. They are the geometry of probability in quantum space.

Diagram 6.5 – Atomic Orbitals as Probability Clouds

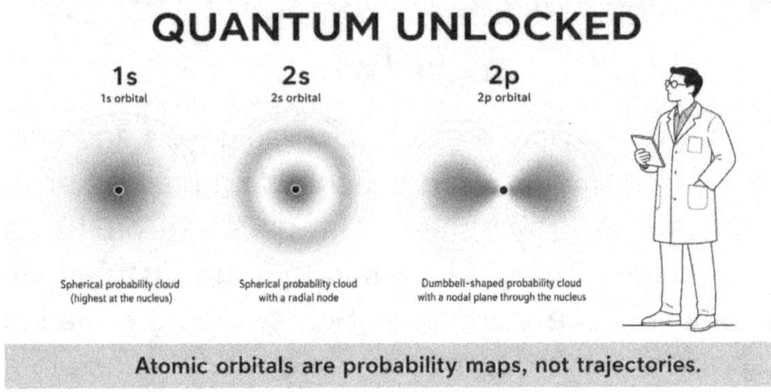


Here is where the analogy to a weather forecast revisits us, now in a richer form. The orbital shape is nature's forecast for where the electron will be found. But unlike a storm forecast, there is no underlying truth being hidden. The electron does not have a secret position within the cloud. The cloud is the most complete description of where the electron is. When you make a measurement, you find the electron at a specific point, but that point was not predetermined. Before the measurement, the electron was genuinely distributed across the cloud, not hiding at a secret spot within it.

## Schrodinger's Equation: The Time Machine for Possibilities

The wave function at a single moment in time is a snapshot of a quantum system's possibilities. But quantum systems do not stay frozen. Atoms absorb and emit light. Electrons change energy levels. Molecules vibrate and rotate. How does the wave function change over time?

This is what Schrodinger's equation governs. The equation is a recipe: given the wave function right now, and given the forces and constraints acting on the system, the equation tells you precisely what the wave function will look like at any future moment. In this sense, Schrodinger's equation is a time-evolution machine for possibilities. It takes a probability landscape and rolls it forward, showing how probabilities shift and flow over time.

To avoid mathematics entirely, think of it this way. Imagine you are tracking a ball of dough as it is kneaded. The dough is not at a single fixed location: it is spread out, stretched, and folded. A recipe tells you how to transform it step by step. Schrodinger's equation is the recipe for the wave function. The wave function is the dough: a spread-out, flexible description of possibility. The equation evolves it forward in time. Each moment, the probability landscape reshapes itself according to strict rules, and those rules predict the dough's configuration at any later time with perfect precision.

Diagram 6.6 – Schrodinger's Equation as a Time-Evolution Machine

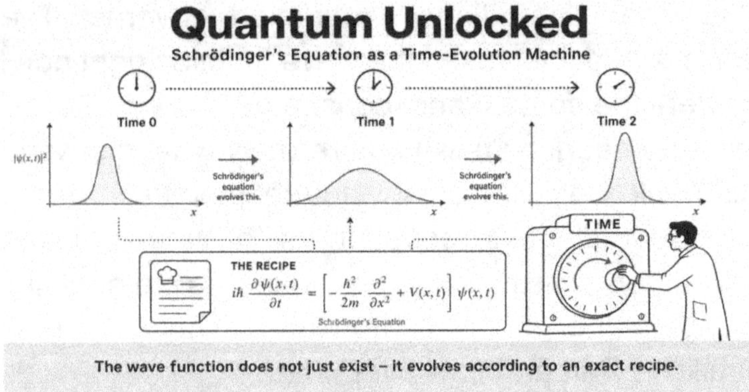


One of the most important properties of Schrodinger evolution is that it is deterministic and reversible. Given the wave function at one moment, the equation uniquely determines the wave function at all future and past moments. There is no randomness in the evolution itself. The randomness in quantum mechanics arises only at the moment of measurement, when a definite outcome is selected from the spread of possibilities. Between measurements, the wave function evolves in a perfectly smooth, predictable way.

## Energy Levels as Standing Waves

Schrodinger's equation predicts something that classical physics could not explain: why atoms have discrete energy levels. When you solve the equation for an electron trapped inside an atom, you find that only certain wave function shapes are stable solutions. These stable solutions are the energy eigenstates, and each corresponds to a specific energy level. Shapes that do not fit the boundary conditions of the atom are not valid solutions.

This is analogous to a guitar string. When you pluck a string fixed at both ends, only certain wavelengths produce stable standing waves: the fundamental frequency, the first harmonic, the second harmonic, and so on. Wavelengths that do not fit the length of the string produce only noise. Schrodinger's equation imposes similar boundary conditions on the electron's wave function inside an atom. Only certain shapes work, and those shapes correspond to the discrete energy levels that produce the spectral lines described in Chapter 3. The discreteness of energy is not an arbitrary rule imposed from outside. It is a natural consequence of confining a wave.

Diagram 6.7 – Energy Levels as Standing Wave Solutions

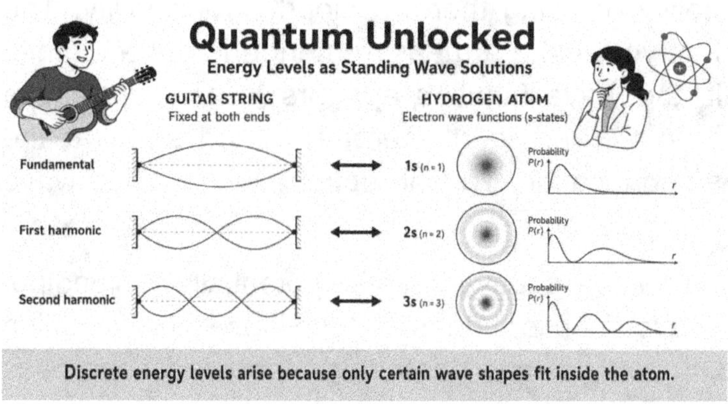


The analogy does reach a limit. A guitar string is a real physical object, and its standing wave patterns are real physical vibrations in the string. The electron's wave function is not a physical vibration in any medium. It is a mathematical object that lives in an abstract space, and its solutions only correspond to measurable energies through Born's rule. Do not let the analogy convince you that the electron is literally wiggling. The analogy captures the mathematical structure of standing waves, not the physical substance.

## Wave Function Collapse: When Possibility Becomes Fact

Here is the part that has caused more debate among physicists than perhaps any other single idea in science. When you measure a quantum system, something dramatic happens to the wave function. Before measurement, the wave function is spread out: the electron has some probability of being here, some probability of being there, a cloud of possibility. The moment you make a measurement and get a definite result, the wave function appears to snap from that spread-out cloud to a sharp, localized state at the measured location. This abrupt change is called wave function collapse.

The name is evocative but also potentially misleading. The wave function does not collapse like a falling building, with pieces scattering in all directions. It is more like a forecast that suddenly becomes a report of actual conditions. One moment, you have a probability distribution spanning many outcomes. The next moment, you have a definite result, and the wave function updates to reflect that result. The question that has consumed physicists for decades is: what exactly causes this transition, and what does it mean physically?

To appreciate the strangeness, consider what collapse implies. If the electron were really spread across multiple positions before measurement, then the act of measurement would somehow reach into the physical world and select one outcome from many. That selection is not deterministic: Born's rule gives you the probabilities, but it cannot tell you in advance which specific outcome will occur. Every time you repeat the measurement on an identically prepared system, you will get the same probability distribution of outcomes, but each measurement result is irreducibly random. This is not a limitation of our instruments. It is the deepest feature of quantum mechanics.

Diagram 6.8 – Wave Function Collapse: From Cloud to Point

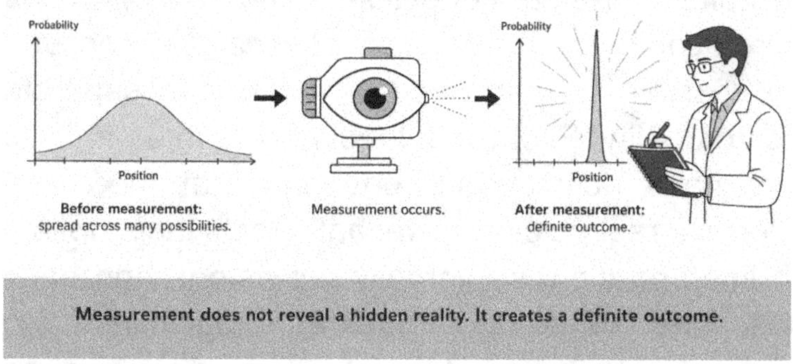

## The Measurement Problem

This brings us to one of the genuine open questions in physics: the measurement problem. Schrodinger's equation describes the smooth, continuous, reversible evolution of the wave function. But wave function collapse is none of those things. It is abrupt, discontinuous, and irreversible. The two processes, evolution and collapse, are governed by different rules, and quantum mechanics itself does not tell us precisely when one applies and when the other takes over.

What counts as a measurement? Is it when a particle interacts with a detector? Is it when the detector signal is amplified into a classical electrical pulse? Is it when a human observer reads the result? Different interpretations of quantum mechanics answer these questions in very different ways, and none is universally accepted. The Copenhagen interpretation, the most widely taught version, says that the wave function is a tool for calculating probabilities and that asking what "really happens" during collapse is not a meaningful physical question. The many-worlds interpretation holds that collapse never actually occurs: every possible outcome occurs, each in a separate branch of reality. Pilot wave theory holds that particles always have definite positions and that the wave function is a real guiding wave that directs their motion. Each interpretation is internally consistent, and each makes the same experimental predictions. No experiment conducted to date has distinguished between them.

This is not a failure of quantum mechanics. The theory is spectacularly successful at predicting experimental outcomes. The measurement problem is a question about what the theory means, not a question about whether its predictions are correct. As an investigator, you should recognize that science can produce a perfectly reliable predictive framework while leaving the deepest question, "what is really going on?", open for continued investigation.

Diagram 6.9 – Interpretations of Wave Function Collapse

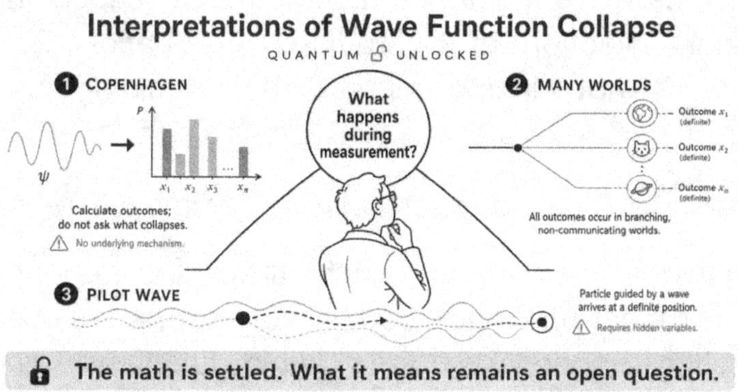


## Atomic Orbitals: The Architecture of Chemistry

Wave functions are not just philosophical puzzles. They are the physical basis for all of chemistry. The shapes of atomic orbitals, determined by solving Schrodinger's equation for electrons bound to nuclei, govern exactly which atoms can bond to which other atoms, what angles chemical bonds form, and how molecules behave. Chemistry is, at its foundation, applied wave function mathematics.

The simplest case is the hydrogen atom, which has one electron and one proton. Its orbitals form a nested hierarchy: the 1s orbital is a sphere, lowest in energy. Above it sit the 2s and 2p orbitals, higher in energy and more complex in shape. Above those come the 3s, 3p, and 3d orbitals, higher still. Each shell represents a range of wave function solutions with different energies. When an electron occupies one of these orbitals, its wave function matches the corresponding stationary solution of Schrödinger's equation. The electron is not circling a track. It is residing in a probability cloud of a specific shape and energy.

## Molecular Bonding and Overlapping Clouds

When two atoms approach each other closely enough that their electron clouds begin to overlap, something interesting happens to their wave functions. The electron wave functions of each atom can combine. Just as two water waves combine on a pond, if they combine constructively, the probability of finding an electron in the region between the two nuclei increases. That concentrated electron density between the nuclei pulls both nuclei toward it, holding them together. This is a covalent chemical bond.

The strength and geometry of a chemical bond are determined entirely by the shapes of the overlapping wave functions. An s-orbital on one atom and a p-orbital on another will overlap in a specific way that produces a specific bond angle and strength. The extraordinary diversity of molecular structures, from simple water molecules to complex proteins and DNA strands, arises from the geometrical rules governing wave-function overlap. When a biochemist describes the shape of a protein's active site, they are ultimately describing the outcome of wave function overlap rules applied to hundreds of atoms. The chemistry of life is quantum mechanics at work.

Diagram 6.10 – Molecular Bonding as Wave Function Overlap

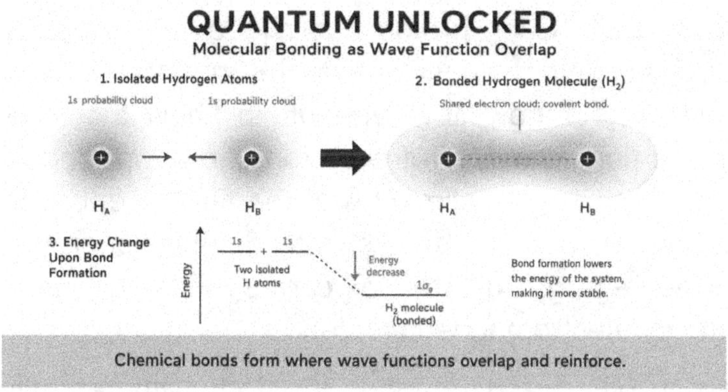


This is also why molecular geometry is not arbitrary. The nitrogen atom in an ammonia molecule forms bonds at angles of about 107 degrees, not 90 or 180 degrees. The oxygen atom in water forms two hydrogen bonds at about 104.5 degrees. These precise angles are predictions of quantum mechanics, arising from the specific shapes of the atomic wave functions involved in bonding. Before quantum mechanics, chemists knew these angles from experiment but had no explanation for them. Schrodinger's equation explained first principles.

## Wave Functions in Medicine: The MRI Machine

The wave function is not only the foundation of theoretical chemistry. It also underpins one of the most important diagnostic tools in modern medicine: magnetic resonance imaging (MRI).

Inside a strong magnetic field, the nuclei of hydrogen atoms, which are single protons, behave like tiny quantum tops. Each proton has a quantum property called spin, which you will encounter in more detail in Chapter 9. For now, it is enough to know that in a magnetic field, the spin of a proton can adopt one of two quantum states: aligned with the field or opposed to it. These two states have slightly different energies, and the energy difference corresponds to radio frequency electromagnetic radiation.

When an MRI machine floods the body with carefully tuned radio waves, protons absorb the energy and flip their spin states. When the radio waves are switched off, the protons relax back to their lower-energy state and emit radio waves of a specific frequency. These emitted signals are detected by the MRI scanner and used to construct a three-dimensional image of the body's internal structure. The contrast in the image, the difference between muscle, fat, fluid, and bone, arises from the fact that protons in different chemical environments relax at different rates. Their wave functions, shaped by the molecular surroundings, produce slightly different emission frequencies. An MRI machine, in essence, reads wave-function information from inside your body and converts it into a diagnostic image.

Diagram 6.11 – MRI: Reading Wave Functions Inside the Body

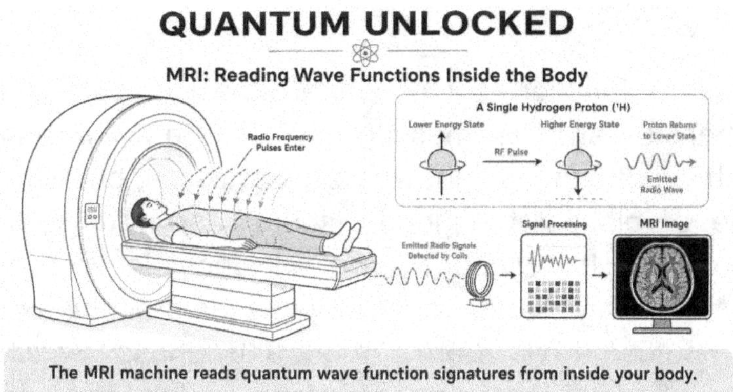


This is not a distant or theoretical application. MRI has become the dominant diagnostic tool for conditions ranging from brain tumors to torn ligaments precisely because it is non-invasive and extraordinarily sensitive. It works because hydrogen atoms are everywhere in the human body (water molecules alone make up about 60 percent of body weight), and because those hydrogen nuclei obey quantum mechanical rules that respond predictably to magnetic and radio frequency fields. Every MRI scan demonstrates that quantum mechanics is not only true but also medically practical.

The wave function of a proton inside your tissue is a real physical object with real consequences. It determines how a proton interacts with the MRI machine's magnetic field. It encodes information about the proton's chemical environment. And it collapses, in the specific quantum sense, every time the scanner's detectors register an emitted signal. The ghost in the machine is at work in your local hospital every day.

## Takeaway

The wave function is the central object of quantum mechanics, and its message is both precise and unsettling. It tells you not where a particle is, but what range of places it might be found and with what probability. It encodes two distinct layers of information: amplitude, which governs the probability of each outcome via Born's rule, and phase, which governs how different paths interfere, reinforcing some possibilities and canceling others.

Schrödinger's equation is the time-evolution machine that rolls this probability landscape forward moment by moment, deterministically and perfectly, between measurements. But when a measurement occurs, the spread-out wave function yields a single definite result, selected at random according to Born's rule probabilities. This transition, called wave function collapse, is one of the deepest and still-debated mysteries in all of physics. The mathematics works perfectly. The meaning remains open.

Atomic orbitals are not orbits. They are three-dimensional probability clouds derived from solving Schrödinger's equation, and their overlapping shapes form the physical basis of every chemical bond in the universe. The angles of chemical bonds, the reactivity of atoms, and the structure of DNA all trace back to the geometry of wave functions. And when a doctor orders an MRI, the machine exploits quantum spin states and wave function transitions in hydrogen nuclei to build images of your internal anatomy. The ghost in the machine is not a ghost at all. It is a precisely defined probability structure that underlies the physical world at every scale, from the nucleus of a hydrogen atom to the chemistry of life to the instruments of modern medicine.

Whether the wave function is a description of physical reality or merely our knowledge of it, the experiments do not yet decide. But what the experiments do confirm, again and again to extraordinary precision, is that Born's rule works, Schrodinger's equation works, and the probability clouds predicted by quantum mechanics are the actual shapes of electron distributions in atoms. You are made of probability clouds. They happen to be very stable ones.

# 7 Spooky Action: Entanglement and the End of Local Reality

Imagine a pair of gloves. You pack one in a box and ship it to a friend across the world. When they open it and find a left-handed glove, you instantly know yours is right-handed. No mystery. The gloves were always a matched pair. You did not look until now.

Einstein believed quantum mechanics worked the same way. When two particles emerge from a shared event, he argued, their properties are fixed at birth, just as the gloves are. We do not yet know the values. Look at one particle, and you learn the other's value, but only because the information was already baked in. Nothing traveled between them. Nothing strange happened.

He was wrong. And the wrongness, confirmed by some of the most carefully designed experiments in the history of physics, turns out to be one of the deepest results science has ever produced. The world is not like the gloves. When two particles are entangled, their properties are genuinely undetermined until measurement. The act of measuring one particle does not reveal a pre-existing value. It participates in creating the outcome. And whatever it creates, the other particle's outcome is instantly correlated, no matter how far apart they are.

Einstein called this spooky action at a distance. He meant it as a criticism. It turned out to be a description.

## Why It Matters

Quantum entanglement is not a curiosity confined to laboratory tabletops. It underlies the emerging technologies of quantum cryptography and quantum communication. It has forced physicists and philosophers alike to revise long-held assumptions about what it means for two objects to be separate, about whether physical properties exist before they are observed, and about the very structure of reality at its most fundamental level.

This chapter is the story of that revision. You will follow the argument from Einstein's original objection in 1935 through the theorem that made it testable in 1964, then through the experiments that settled the debate in the 1980s and beyond. By the end, you will understand not only what entanglement is but what it implies, and you will be equipped to think clearly about both the genuine strangeness and the very real limits of what it can do.

## The Classical Comfort of Separated Objects

### What Local Realism Assumed

Before quantum mechanics, physics operated under two assumptions so intuitive that most people never noticed them as such. The first is realism: the idea that physical properties exist whether or not anyone is looking at them. A particle has a definite position, a definite spin, and a definite momentum. These values are real features of the world. They are not created by measurement; measurement merely reveals them.

The second assumption is locality: the idea that what happens at one location cannot instantly influence what happens at another. If you want to affect something far away, you need to send a signal, and signals cannot travel faster than light. This is not merely a practical limit—Einstein's theory of special relativity built locality into the foundations of physics. Cause precedes effect, and causation requires a signal that travels through space at a finite speed.

Together, local realism and classical physics form the bedrock. Objects have definite properties. Distant objects do not influence each other without an intermediary. The universe behaves like a collection of separate, self-contained things interacting only through direct contact or through fields that carry influence at finite speed.

Quantum mechanics attacked both assumptions simultaneously, and in 1935, Albert Einstein decided it had gone too far.

Diagram 7.1 - Classical Locality: The Separated Objects Model

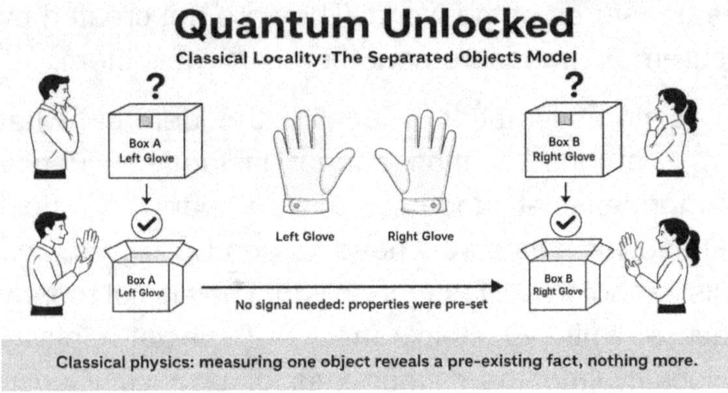


## The Gloves Analogy and Its Limits

The glove scenario captures the classical picture perfectly. You separate a matched pair. You ship them to opposite ends of the planet. The instant your friend opens their box and sees a left-handed glove, they know yours is right-handed. The correlation is perfect. The information is instantaneous. And yet nothing traveled between the gloves. The result was determined when you packed the boxes.

This is the hidden variable scenario. The property, in this case, handedness, was fixed in advance. The measurement read out a pre-existing fact. Any theory that works this way, with all correlations explainable by prior shared information, is called a local hidden variable theory. The word local means the explanation does not require any instantaneous influence between the distant objects. The word "hidden" means that the variables that determine the outcomes may not be accessible to us, but they exist.

For decades, hidden variable theories seemed perfectly plausible. The correlations predicted by quantum mechanics could, in principle, be explained this way. The glove story was not just an analogy. It was a coherent alternative interpretation of the experimental data. Then John Bell found a way to test it.

But before Bell, there was Einstein, and before Bell's test, there was a thought experiment that sharpened the question until it could no longer be avoided.

## The EPR Thought Experiment

### Einstein, Podolsky, and Rosen

In 1935, Albert Einstein, Boris Podolsky, and Nathan Rosen published a paper that became known by their initials: EPR. The paper was not a celebration of quantum mechanics. It was an attack. Its goal was to show that, however accurate its predictions, quantum mechanics was an incomplete description of physical reality.

The argument ran as follows. Imagine you prepare two particles so that their total momentum is known and their total position relationship is fixed. Quantum mechanics says that neither particle has a definite momentum or position on its own; only a shared state exists. Now separate the particles by a large distance. Measure the position of particle A. You now know the position of particle B exactly, without ever touching it. Measure the momentum of particle A instead. You now know the momentum of particle B exactly, without ever touching it.

Einstein's point was this: if measuring particle A gives you information about particle B without any physical interaction with particle B, then particle B must have had a definite position and a definite momentum all along. Quantum mechanics failed to represent those definite values. The wave function was incomplete. The real theory, the complete theory, would include hidden variables that described the actual properties of each particle.

Einstein called this argument a demonstration of the incompleteness of quantum mechanics. He was not claiming that quantum mechanics was wrong in its predictions. He was claiming that the wave function, like a statistical description of a gas, was accurate on average but hiding a deeper, fully determined reality underneath.

Diagram 7.2 - The EPR Thought Experiment Setup

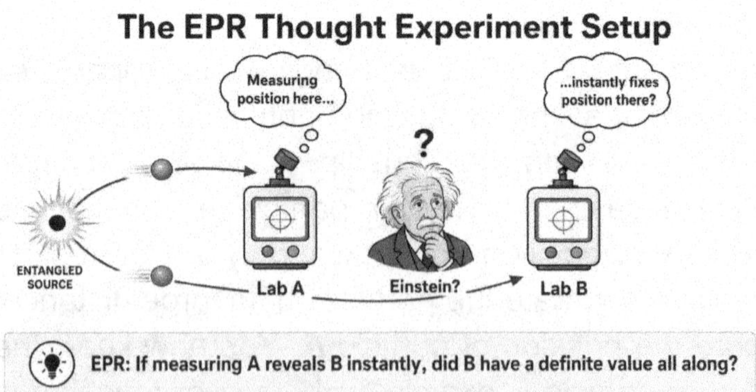

## Bohr's Reply

Niels Bohr replied to EPR within months. Bohr's position, the core of what became the Copenhagen Interpretation, was that it is meaningless to speak of a particle's position or momentum before a measurement is made in a context that can reveal it. The measurement and the property are inseparable.

According to Bohr, Einstein's argument failed because it assumed that particle B had both a definite position and a definite momentum simultaneously. But position and momentum cannot be simultaneously defined. The measurement context determines which property is real, and asking about both at once imports classical assumptions that quantum mechanics does not support.

The exchange between Einstein and Bohr was one of the most productive arguments in the history of physics. Neither man had a way to definitively test which picture of reality was right. That would require another three decades and a young Irish physicist working at CERN.

Diagram 7.3 - Einstein vs. Bohr: Two Views of Quantum Reality

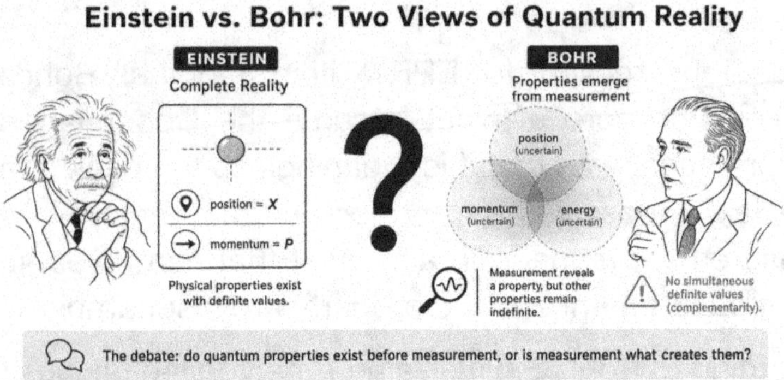


# Hidden Variable Theories

## The Appeal of a Deeper Theory

Einstein was not alone in his discomfort. Several physicists attempted to construct hidden-variable theories: frameworks in which every quantum particle carries a complete set of definite properties, with the quantum wave function representing only our ignorance of those properties. The most famous attempt was by David Bohm in 1952.

Bohm's pilot wave theory was a genuine achievement. He showed that it is mathematically possible to describe quantum mechanics using a classical particle guided by a real wave. The wave evolves according to the Schrodinger equation. The particle rides the wave, following its contours. The apparent randomness of quantum measurement is not fundamental. It is the result of our ignorance of the particle's exact starting position in the wave.

Bohm's theory reproduced every prediction of standard quantum mechanics. It was deterministic. It assigned definite positions and momenta to particles at all times. Einstein read it with interest. But even Bohm's theory had a troubling feature: to reproduce the correlations between entangled particles, the pilot wave had to be nonlocal. A change to the wave at particle A's location had to instantly affect the wave at particle B's location, no matter how far apart they were. Bohm's theory avoided randomness but not nonlocality.

This raised a sharper question: Is nonlocality avoidable? Or is it built into any theory that reproduces quantum mechanics' predictions? John Bell answered that question in 1964.

Diagram 7.4 - Bohm's Pilot Wave: Particle Riding a Guiding Wave

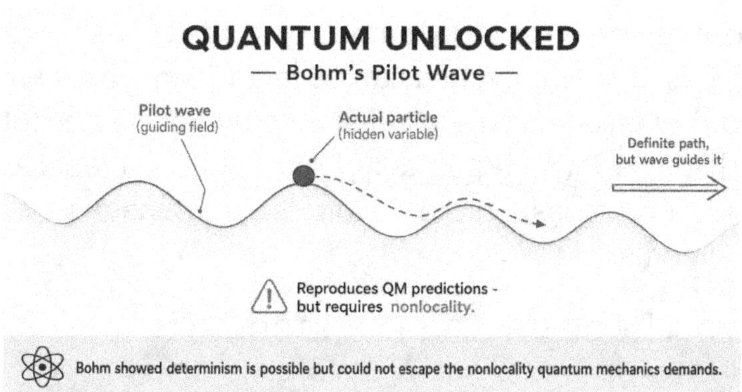

### Why Hidden Variables Are Attractive

The appeal of hidden variable theories runs deeper than physics. It connects to a basic human intuition about the world: that things are what they are, independent of our looking. The idea that an electron genuinely has no spin direction until the moment we measure it seems to violate something fundamental about what it means for reality to exist.

This intuition is so strong that even after decades of experimental evidence, some physicists continue to search for deterministic, locally causal accounts of quantum phenomena. The effort is motivated by a legitimate philosophical concern: science should describe a world that would exist whether or not anyone was around to observe it.

What Bell showed in 1964 was that this intuition, combined with the requirement of locality, makes a quantitative prediction that can be tested. And the tests failed. Not failed in a marginal or ambiguous sense, but failed decisively, repeatedly, and with increasing precision across more than five decades of experiments.

## Bell's Theorem and the Inequality

### Setting Up the Test

John Stewart Bell was born in Belfast in 1928 and worked as a particle physicist at CERN, with deep interests in the foundations of quantum theory. In a 1964 paper, he proved a theorem that transformed the

Einstein-Bohr debate from a philosophical argument into an experimental question.

Bell's key insight was this: if local hidden variables exist, correlations between measurements on separated particles must obey a mathematical constraint. He derived this constraint as an inequality. If the measured correlations satisfy the inequality, local hidden variables could account for them. If the measured correlations violate the inequality, no local hidden variable theory can explain them. The quantum world itself would have to be nonlocal.

Bell's theorem applies to any theory in which properties carried locally by the particles determine outcomes. It assumes nothing about the specific nature of those properties. The conclusion is stark: local realism and quantum mechanics cannot both be correct.

Diagram 7.5 - Bell's Inequality: Classical Prediction vs. Quantum Prediction

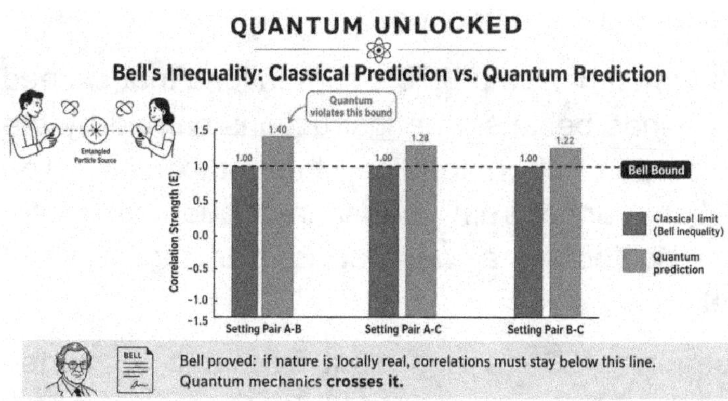

## The Logic of the Inequality

To understand Bell's inequality, it helps to think about a stylized version of the experiment. Two particles, call them A and B, are produced together and sent in opposite directions to two detectors. Each detector can be oriented to measure the particle's spin along any of three directions: 0 degrees, 60 degrees, or 120 degrees.

For each pair, the two detectors independently choose a direction. The result at each point is either +1 or -1, spin-up or spin-down. The experiment is repeated thousands of times with randomly chosen, independent settings.

If local hidden variables determine outcomes, the correlations must satisfy Bell's inequality. The intuition is that the particles carry a fixed set of instructions, like a prearranged cheat sheet, that determine outcomes at each detector. The correlations any such instruction set can produce are bounded.

Quantum mechanics predicts correlations that exceed those bounds because the instructions carried by the particles do not determine the outcomes. The correlations arise from the shared quantum state, which is nonlocal in a way that no instruction set can replicate.

The beauty of Bell's argument is that it is model-independent. You do not need to specify a mechanism or know the hidden variables. You measure the correlations and compare them to Bell's bound. The bound is either satisfied or violated, and the answer determines whether local hidden variables can explain the world.

Diagram 7.6 - How Bell's Experiment Works: Two Detectors, Three Settings

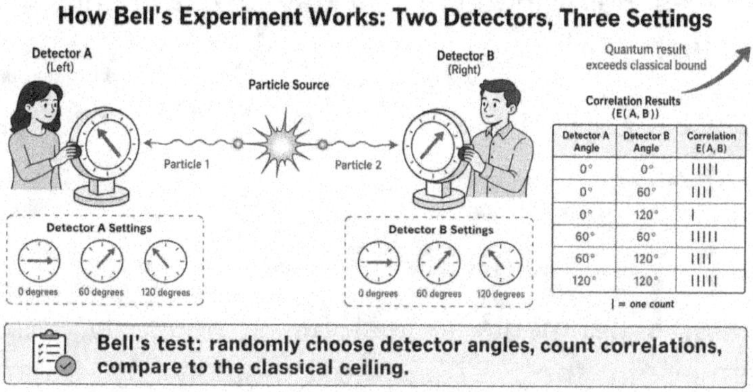


## What Bell Actually Proved

It is worth pausing to appreciate what Bell's theorem actually establishes. It does not say that quantum mechanics is correct. It says that if quantum mechanics is correct, then nature is nonlocal in the following sense: the outcomes of measurements on two widely separated particles cannot be explained by any theory in which each particle carries a set of locally defined properties fixed at the moment of creation.

This is not merely saying that quantum mechanics is a good predictive tool. It is saying something about the structure of physical reality. Either some influence connects the two particles at the moment of measurement, or the correlations arise from a holistic shared state that cannot be decomposed into properties of the individual particles alone.

The phrase 'spooky action at a distance' now has a precise technical meaning. It violates Bell's inequality. And it was Bell's theorem that gave experimenters a way to test for it.

## Aspect's Experiments and the Closing of Loopholes

### The First Tests

The first experiments to test Bell's inequality were performed in the 1970s, most notably by Stuart Freedman and John Clauser at Berkeley. These experiments produced correlations that violated Bell's inequality, consistent with the predictions of quantum mechanics. But the experiments had what physicists call loopholes: ways in which a local hidden-variable theory could, in principle, still produce the observed correlations without genuine nonlocality.

The most important loophole is called the locality, or communication, loophole. If the detector settings are not chosen randomly and independently after the particles are already in flight, then there is a window of time in which a signal traveling at the speed of light could carry information from one detector to the other. The particles, or the detectors, might conspire to produce the observed correlations without any genuine instantaneous influence.

A second important loophole is the detection loophole. Real detectors do not register every particle. If detected particles are not a fair sample of all produced particles, a local hidden-variable theory might exploit that sampling bias to mimic quantum correlations.

Diagram 7.7 - The Locality Loophole: Why Early Tests Left Room for Doubt

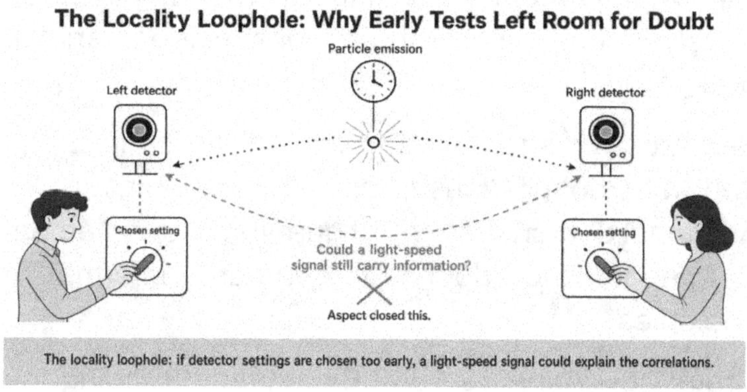


## Alain Aspect's Breakthrough

In 1981 and 1982, French physicist Alain Aspect and his team at the University of Paris addressed the locality loophole directly. Their key innovation was to switch the detector settings after the particles were already in flight, fast enough that no light-speed signal could travel from one detector to the other in time to influence the outcome.

Aspect's experiments used pairs of photons produced by calcium atoms decaying in two steps, with each photon emitted in opposite directions and correlated in polarization. The photons were detected at two stations, roughly 10 meters apart, with polarization analyzers randomly switched as the photons traveled, closing the communication window.

The results were unambiguous. The correlations violated Bell's inequality by several standard deviations. The violation was not marginal. It was exactly what quantum mechanics predicted, and it was decisively incompatible with any local hidden variable theory.

Aspect received the Nobel Prize in Physics in 2022, shared with John Clauser and Anton Zeilinger, for foundational demonstrations of quantum entanglement that opened the door to quantum information technologies.

Diagram 7.8 - Aspect's 1982 Experiment: Closing the Locality Loophole

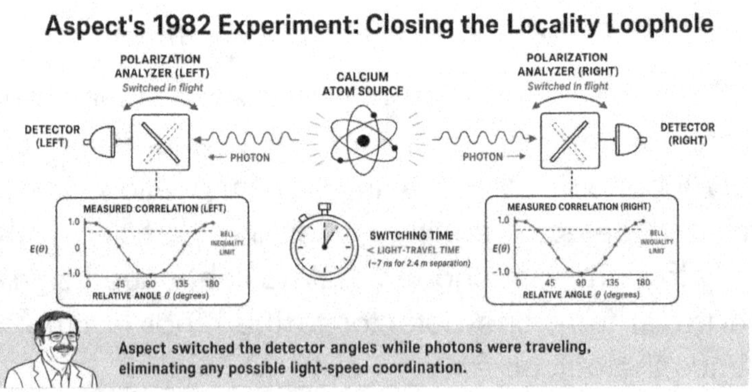

## Loophole-Free Tests and the Modern Confirmation

Aspect's experiments were a landmark, but they left minor loopholes. The detection loophole required detector efficiencies much higher than early technology could achieve. For the next three decades, physicists worked to close each remaining loophole, steadily improving detector technology and experimental design.

In 2015, a team at Delft University of Technology performed what is widely regarded as the first genuinely loophole-free Bell test. Using nitrogen-vacancy defects in diamonds separated by 1.3 kilometers, they closed both the locality loophole and the detection loophole simultaneously. The result: Bell's inequality was violated, and local hidden variables were definitively ruled out.

Simultaneously, teams at NIST in Boulder and at the University of Vienna performed independent loophole-free Bell tests. All three 2015 experiments agreed: the universe does not run on local hidden variables.

A still more demanding test came in 2018: the Big Bell Test, in which over 100,000 human volunteers generated 97 million random detector-setting choices across 13 laboratories on 5 continents. Bell's inequality was violated in every laboratory, and the results, published in Nature, confirmed that quantum entanglement is genuinely nonlocal.

Diagram 7.9 - The 2015 Loophole-Free Bell Test at Delft University

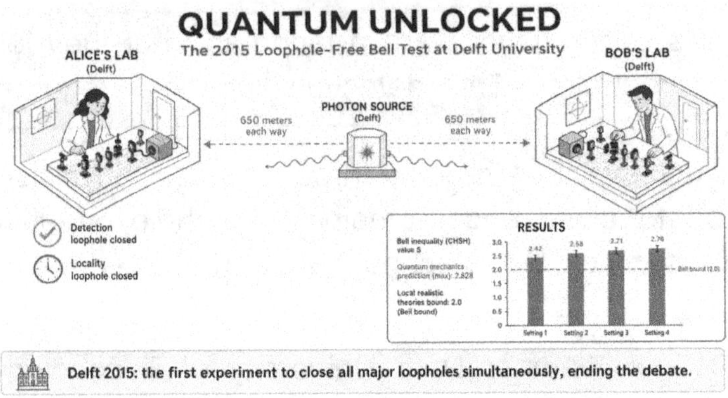


# What Entanglement Is and Is Not

## Not a Communication Channel

The single most common misunderstanding about quantum entanglement is the belief that it can be used to communicate information faster than the speed of light. It cannot. This is not a technological limitation that might be overcome with better engineering. It is a fundamental feature of entanglement.

Here is why. When you measure an entangled particle, the outcome you get is random. You cannot choose to spin up or spin down. The result is inherently unpredictable. The correlation between the two particles is real and nonlocal. But you have no control over your outcome, and neither does your partner. The correlation only becomes visible when you later compare results through an ordinary classical channel.

Imagine you and a friend each hold one particle of an entangled pair. Your friend measures theirs and gets spin-up; yours will measure spin-down. But your friend cannot use this to send a message, because they cannot control the result. They get whatever the quantum dice deliver. You also get a random result, so you cannot tell whether your partner has measured yet. The only way to see the correlation is to compare notes, and that comparison requires a classical signal.

This is sometimes called the no-communication theorem, and it follows directly from the mathematics of quantum mechanics. Entanglement creates correlations that are stronger than any classical explanation can provide, but those correlations are passive. They cannot carry a message.

Diagram 7.10 - Why Entanglement Cannot Carry Information Faster Than Light

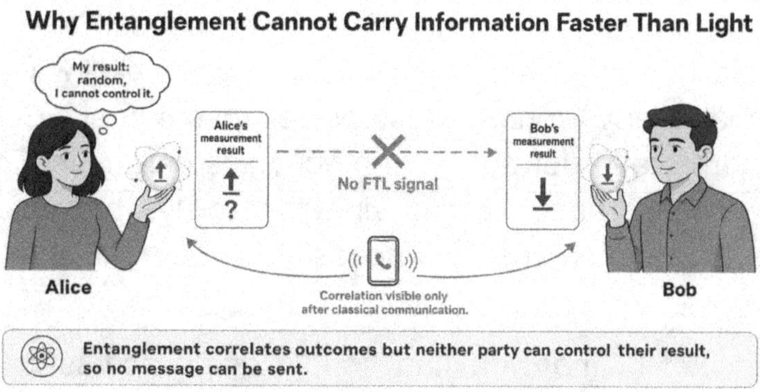

## What Entanglement Actually Tells Us About Reality

If entanglement cannot send a signal, why does it matter so profoundly? Because it tells us something about the nature of physical reality that goes beyond what signals can do.

Bell's theorem, combined with experimental results, establishes that no theory that assigns definite local properties to particles before measurement can reproduce the observed correlations. This means one of three things must be true. Either measurements at one location instantly influence outcomes at another location (nonlocality). Or particles do not have definite properties before measurement (anti-realism)—or both.

Most physicists accept some form of nonlocality, anti-realism, or both. Copenhagen leans toward anti-realism: properties do not exist before measurement. Bohmian mechanics accepts nonlocality: the pilot wave is global. Many Worlds sidesteps collapse entirely. But no interpretation of quantum mechanics can maintain both locality and realism simultaneously, because Bell's theorem proves it is impossible.

This is the genuine content of spooky action at a distance. Not that you can send signals across the universe instantly. But the classical picture of a world made of separate, self-contained things, each with its own local properties, is incompatible with the world we actually live in.

# Quantum Entanglement in Technology

## Quantum Cryptography

Among the practical applications of entanglement, quantum cryptography is the most developed. Quantum key distribution, or QKD, uses the principles of quantum mechanics to create encryption keys that are theoretically unbreakable.

The security of quantum cryptography rests on a physical fact: any attempt to intercept a quantum signal disturbs it in a detectable way. If an eavesdropper tries to read the photons carrying your encryption key, the eavesdropping introduces errors you and your partner can detect. You know you are being watched, and you can discard the compromised key.

In entanglement-based QKD, a source produces entangled photon pairs. Alice receives one photon from each pair, Bob receives the other, and each measures in a randomly chosen setting. Their results are correlated in a way that an eavesdropper cannot replicate without introducing detectable errors. They compare a fraction of measurements over a public channel; if those match quantum predictions, the remaining results form a secure shared key.

Quantum networks using QKD have been demonstrated over hundreds of kilometers of fiber and via satellite. China's Micius satellite, launched in 2016, demonstrated QKD from space to ground stations. Several countries have deployed or are deploying quantum communication networks for secure government and financial use.

## Quantum Teleportation

Quantum teleportation is another real application of entanglement, and it is even more frequently misunderstood than quantum cryptography. The word calls to mind science fiction: a person appearing instantly elsewhere. Quantum teleportation transports neither matter nor energy. What it transfers is quantum information: the precise quantum state of a particle.

Here is how it works. Alice has a particle in an unknown quantum state, and she wants to send it to Bob. Alice and Bob each have one particle of a pre-shared entangled pair. Alice performs a joint measurement on her unknown particle and her half of the entangled pair. This measurement gives her two bits of classical information. She sends those two bits to Bob through an ordinary classical channel. Bob uses those two bits to perform a simple operation on his half of the entangled pair. After Bob's operation, his particle is now in the same quantum state as Alice's unknown particle.

Two things are crucial. First, Alice's original particle is destroyed by the measurement: the quantum state has been moved rather than copied, consistent with the no-cloning theorem. Second, the classical bits must travel to Bob through a conventional channel, so the transfer cannot exceed the speed of light. Quantum teleportation is not faster-than-light communication.

Quantum teleportation has been demonstrated experimentally with photons and ions, over distances ranging from tabletop experiments to more than 1,400 kilometers via satellite. It is a key primitive in quantum networking, enabling the transmission of quantum states between nodes without physically moving particles.

Diagram 7.11 - Quantum Teleportation: Moving a State, Not a Particle

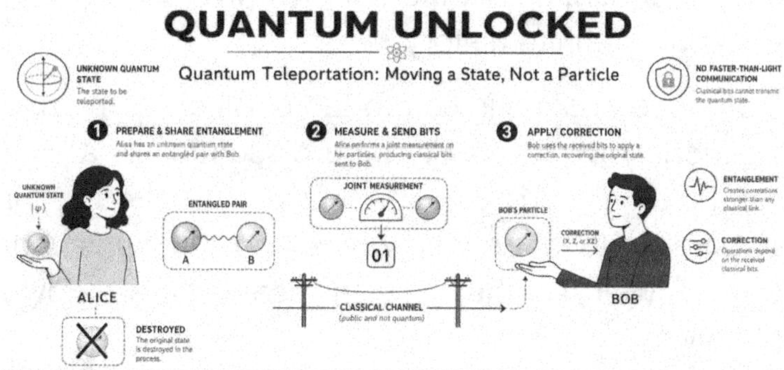


Quantum teleportation moves a quantum state from one place to another but requires a classical channel and destroys the original.

## Implications for Locality and Reality

### The End of Local Realism

The experimental violation of Bell's inequality is not just a technical result. It is a philosophical watershed. For centuries, the scientific worldview rested on the assumption that the universe is made of things: objects with definite properties that interact through local forces. Quantum entanglement, as confirmed by Bell tests, tells us this picture is wrong at the fundamental level.

Exactly what should replace it is still debated. The physics community does not have a single agreed-upon interpretation of quantum mechanics. But the various interpretations share a common starting point: local realism is untenable. Something has to give. Either particles do not have definite properties before measurement, or the properties of distant particles are not independent of each other, or the universe does not have a single definite history.

None of these options is comfortable. All of them have been seriously explored. And the discomfort is a sign of genuine progress: we have pushed our understanding to a frontier where the intuitions that served us well at the human scale break down completely.

## The Nonseparability of the Quantum World

Physicists use the word nonseparability to describe the feature of entangled systems that Bell's theorem put beyond doubt. A classically composed system is separable: you can always describe it by describing each of its parts individually and then combining those descriptions. An entangled quantum system is not separable in this sense. Its complete description cannot be factored into independent descriptions of the individual particles.

This nonseparability is not a limitation of our instruments. It is a feature of the quantum state itself.

Two entangled particles, no matter how far apart, are described by a single joint state. Measuring one is, in a mathematically precise sense, an act that is not confined to one location. The quantum state responds globally.

This has led some physicists and philosophers to argue that the fundamental furniture of the world is not particles but quantum states defined over all of space. Individual particles are features of a holistic description, not independent entities. This is a radical conclusion, but it follows directly from taking Bell's theorem and the experimental results seriously.

## What This Does Not Mean

It is equally important to be clear about what entanglement and nonlocality do not imply. They do not mean that consciousness affects experimental outcomes. They do not mean that distant objects are secretly connected through some medium that could carry psychic signals. They do not make the universe fundamentally mental rather than physical.

Quantum nonlocality is a precise technical claim about correlations between measurement outcomes. It violates classical locality in a specific, mathematically characterized way. The violation is real and experimentally confirmed. But it does not license arbitrary claims about distant influences. The correlations are passive: they cannot carry messages, transmit energy, or violate causality.

The universe is stranger than the classical picture. But it is strange in a specific, well-defined, experimentally measured way. That is what makes the strangeness scientific rather than mystical.

## Takeaway

You began this chapter with a pair of gloves and ended with a revolution in our understanding of physical reality. The journey took you from Einstein's intuition that quantum mechanics must be incomplete, through Bohr's defense of a world in which properties do not exist before measurement, through Bell's theorem that made the dispute testable, and through the experiments that settled it.

What the evidence shows is this: the correlations between entangled particles are stronger than any local hidden variable theory can produce. No pre-arranged instruction set carried by the particles can account for them. The quantum world is nonlocal in the technical sense that distant measurement outcomes are correlated in ways that shared local properties cannot explain. Aspect confirmed this in 1982, the Delft experiment closed all loopholes in 2015, and the global Big Bell Test confirmed it again across 13 laboratories on 5 continents.

Entanglement does not allow faster-than-light communication. The correlations are passive. The outcomes are random at each detector. You can only see the nonlocal correlation after comparing notes through a classical channel. But the correlation itself is real, not a statistical artifact, not a failure of imagination, but a feature of nature that any complete theory must account for.

What this means for reality depends on which interpretation of quantum mechanics you find most compelling, and that question remains open. But the experimental result is not open. Local realism is not how the world works. The universe is not a collection of separate objects with separate histories. At the quantum level, separation is not what it seems. Two particles, created together and measured light-years apart, remain linked in a way that has no classical counterpart.

Einstein called it spooky. He was right about the name and wrong about the conclusion. The spookiness is real, measured, and one of the deepest truths physics has ever produced.

# 8 Through the Wall: Quantum Tunneling and the Physics of the Impossible

Picture a ball rolling across a flat surface toward a hill. You know, without calculating anything, that if the ball does not have enough kinetic energy to reach the top, it will slow, stop somewhere on the slope, and roll back the way it came. That is one of the most intuitive facts in all of physics. Energy in equals energy out. A barrier is a barrier. The ball cannot pass through solid ground. It can only go over or come back.

Now imagine watching that ball roll toward the hill, slow as expected, and then, without ever appearing at the top, materializing on the far side as if the hill were not there at all—no trick photography. No tunnel was dug underneath. The ball appears on the other side, having passed through a region it never had enough energy to enter. This is not science fiction. This is a measurable, repeatable, technologically exploited physical fact. It is called quantum tunneling, and once you understand it, you will see it everywhere: in the heart of the sun, in radioactive atoms, in the flash memory chip inside every smartphone on the planet.

Why It Matters: Quantum tunneling is not a curiosity at the fringe of physics. It is a central mechanism of nature. Without tunneling, nuclear fusion in stars would be millions of times slower, and the sun would have burned out long before life had a chance to evolve.

Without tunneling, certain radioactive elements would never decay, and the geological heat that keeps Earth's interior molten would be dramatically different. Without tunneling, there would be no scanning tunneling microscope, no tunnel diode, no flash memory. Tunneling is not a rare exception to the rules. In the quantum world, it is one of the rules.

## The Classical Verdict: Why Tunneling Should Be Impossible

To appreciate why quantum tunneling is so astonishing, you need to understand the classical picture it violates. In classical mechanics, the energy of a particle is the sum of two contributions: its kinetic energy, which is the energy of its motion, and its potential energy, which is the energy stored in its position relative to forces acting on it. A ball at the top of a hill has high potential energy and (momentarily, at the peak) low kinetic energy. A ball at the bottom of a valley has low potential energy and, if it has been sliding down, high kinetic energy. The total is conserved.

A potential barrier is a region where the potential energy exceeds the particle's total energy. If a particle approaches such a region, classical mechanics issues a verdict that seems absolute: the particle cannot enter. To enter, it would need kinetic energy to pass through the barrier. But inside the barrier, the particle's potential energy exceeds its total energy. That would require negative kinetic energy. And kinetic energy, being related to the square of velocity, cannot be negative. Classical physics does not allow it. The particle is reflected. The barrier holds. End of story.

This classical verdict holds for objects large enough to see. The ball on the hill really does roll back if it lacks the energy to crest the top. A car approaching a hill that it cannot climb really does coast to a stop and reverse. These are not approximate truths. For macroscopic objects, the classical description is essentially exact. The barrier is as real and solid as it feels.

## Energy, Barriers, and the Classical Cutoff

To make this precise without resorting to equations, imagine a simple diagram. On the left, a particle approaches from a flat region. In the middle, the potential energy rises sharply into a rectangular hump, like a speed bump of pure energy. On the right, the potential energy drops back to its original level. The hump has a definite height and a definite width. The particle has a definite total energy, represented as a horizontal line. If the horizontal line sits below the top of the hump, the particle is classically forbidden from the interior of the barrier. It will bounce back every single time.

This picture is clean, deterministic, and comforting. It matches every instinct we have about the physical world. Objects obey energy bookkeeping. No energy, no passage. The wall is real. Except, of course, the evidence from the quantum world says the case is not closed at all.

Diagram 8.1 – Classical Energy Barrier: Ball Blocked by a Hill

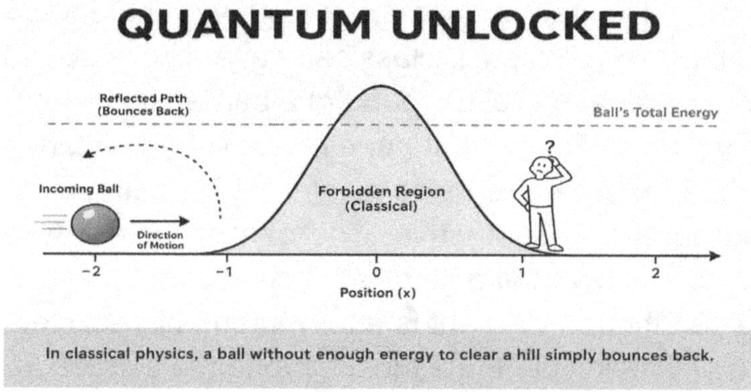


The classical picture also makes a very strong prediction: the outcome is either complete passage or complete reflection, and the deciding factor is a sharp threshold. Either you have enough energy, or you do not. There is no probability involved, no partial transmission, no leakage through the wall. The boundary is crisp and absolute. This prediction works perfectly for baseballs and automobiles. It fails for electrons and alpha particles.

## Wave Functions in Forbidden Territory

The first step toward understanding tunneling is remembering what the previous chapters established about quantum particles. They are not tiny billiard balls following precise trajectories. They are described by wave functions, mathematical objects that encode probability amplitudes across space. Before any measurement is made, the particle does not have a single definite position. It has a spread of possible positions, weighted by the square of the wave function.

This wave-like nature changes the story of the energy barrier entirely. When a classical wave, like a sound wave or a water wave, encounters a barrier, it does not simply stop at the edge. It penetrates a short distance into the barrier before dying away. This penetration is called an evanescent wave. You can observe it with light: shine a beam at a glass surface at a steep angle, and even though the light is totally internally reflected, some of it leaks into the air just beyond the surface for a fraction of a wavelength. The wave does not simply stop at the boundary. It fades.

The quantum wave function behaves the same way when it encounters a potential barrier. Even in a region where a classical particle would be forbidden, the wave function does not drop instantly to zero at the edge of the barrier. It decays exponentially inside the barrier, falling off steeply but not instantaneously. If the barrier is thin enough, the wave function does not reach zero before the far edge. At the far side of the barrier, there is still a nonzero amplitude. And a nonzero amplitude means a nonzero probability of finding the particle there.

Diagram 8.2 – Wave Function Penetrating a Potential Barrier

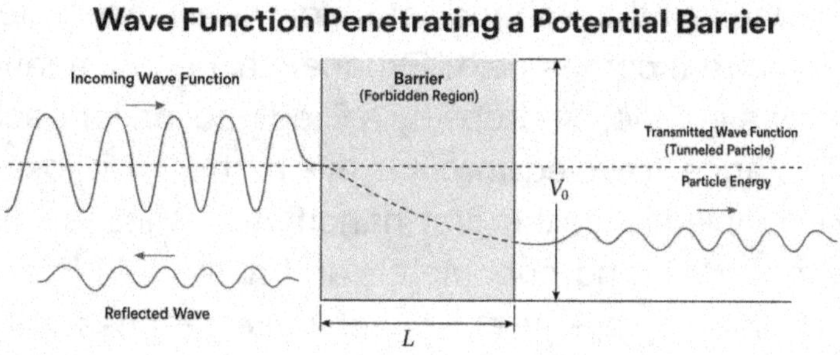


The quantum wave function leaks into forbidden territory and re-emerges on the far side as a tunneled particle.

This is the core of tunneling: the wave function does not simply stop at the wall. It decays inside the wall, and if the wall is thin enough, it emerges on the other side with enough amplitude to give the particle a real probability of appearing there. The particle has not gone over the barrier. It has not gone around. It has passed through a region that was, classically speaking, entirely forbidden to enter.

This is strange enough to deserve a pause. The particle's total energy has not changed. It did not borrow energy to climb over. The theory says this: when the particle is measured on the far side, it is there with a definite, measurable probability, and its total energy remains unchanged. The bookkeeping is preserved. The intermediate steps, however, are not the steps classical mechanics would recognize.

## Exponential Decay Inside the Barrier

The behavior of the wave function inside the barrier is mathematically described by exponential decay, and understanding that shape matters qualitatively for everything that comes next. Exponential decay means that the wave function drops by a fixed fraction for each fixed distance traveled through the barrier. If it loses half its amplitude in the first nanometer, it loses half again in the second nanometer, and half again in the third. It shrinks rapidly but never quite reaches zero until the barrier ends.

This exponential character has a critical implication: the probability of tunneling is extremely sensitive to the barrier width. Double the barrier width, and the wave function that emerges on the other side is not half as large; it is exponentially smaller. Over atomic distances, tunneling is a dominant effect. Over distances we can see with our eyes, the probability drops to a number so vanishingly small that it might as well be zero. The quantum world is allowing something that is technically possible but fantastically improbable for large objects, while being unavoidably probable for small ones.

Diagram 8.3 – Exponential Decay of the Wave Function Inside a Barrier

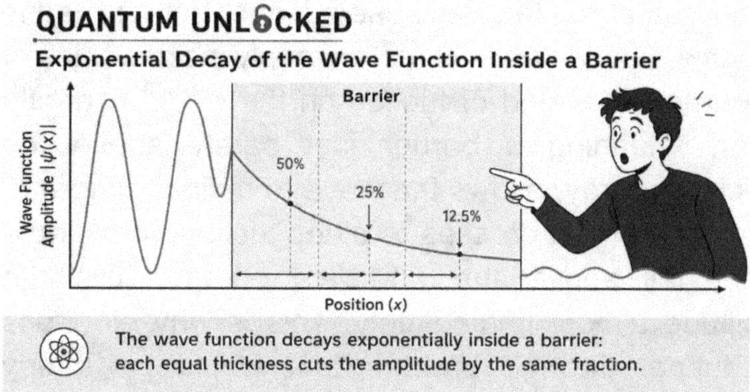


The height of the barrier matters too. A taller barrier, meaning a higher potential energy in the forbidden region, causes the wave function to decay more steeply. The wave loses amplitude faster with each step into the barrier. So for a given thickness, a taller barrier produces a smaller transmitted amplitude and a lower tunneling probability. Both height and width work in the same direction: greater height and width mean less tunneling.

## The Probability of Tunneling: Height, Width, and Mass

Three variables govern the probability of tunneling: the height of the potential barrier above the particle's energy, the barrier width, and the particle's mass. Understanding how each one affects the outcome is essential, and fortunately, each effect is intuitive once you have the wave-function picture in mind.

Barrier height matters because a higher barrier means the wave function decays more steeply inside. Think of the barrier as a kind of friction on the wave. A gentle friction, meaning a barrier just barely above the particle's energy, allows the wave to maintain more of its amplitude as it crosses. A steep friction, meaning a barrier far above the particle's energy, kills the amplitude almost immediately. This is why electrons can tunnel through barriers that would be utterly impenetrable for protons. The proton, being nearly two thousand times heavier than the electron, has a wave function that decays much faster inside a barrier of the same height.

Barrier width is the other dominant factor and follows the same exponential logic. A thin barrier, perhaps a nanometer or two, allows a significant fraction of the wave function's amplitude to survive the crossing. A barrier 10 nanometers thick may reduce the amplitude so drastically that the tunneling probability becomes negligible. Sensitivity to width is one of the most remarkable properties of tunneling, and it is what makes the scanning tunneling microscope possible.

## How Mass Enters the Picture

Mass is the third variable, and it is the one that explains why quantum tunneling is an atomic phenomenon and not a human-scale one. The rate at which the wave function decays inside a barrier depends on both the barrier height and the particle mass. A heavier particle has a shorter quantum wavelength at a given energy. That shorter wavelength means its wave function oscillates more rapidly and decays more steeply inside barriers. For a particle as massive as a grain of dust, the decay inside any macroscopic barrier is so extreme that the tunneling probability is a number so small it would take the lifetime of the universe many times over to observe even one event. For an electron, tunneling through barriers a few nanometers wide is routine, rapid, and technologically useful.

Diagram 8.4 – How Barrier Height, Width, and Mass Affect Tunneling Probability

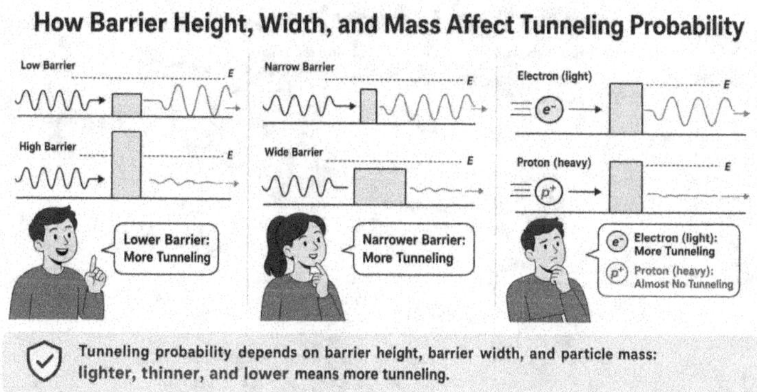


This mass dependence is why tunneling is invisible in everyday life. We do not phase through walls because our mass makes the probability essentially zero. The quantum rules apply to us as much as to electrons. The outcome scales with mass in a way that makes tunneling entirely negligible at human scales.

To summarize the three-variable picture: more tunneling happens when the barrier is shorter above the particle's energy, when the barrier is narrower, and when the particle is lighter. All three factors operate through the same exponential decay of the wave function and push in the same direction. A thin, low barrier traversed by a light particle can have tunneling probabilities approaching one-half or more. A thick, tall barrier traversed by a heavy particle can have probabilities so small they are practically unmeasurable.

Diagram 8.5 – Tunneling Probability as a Function of Barrier Width

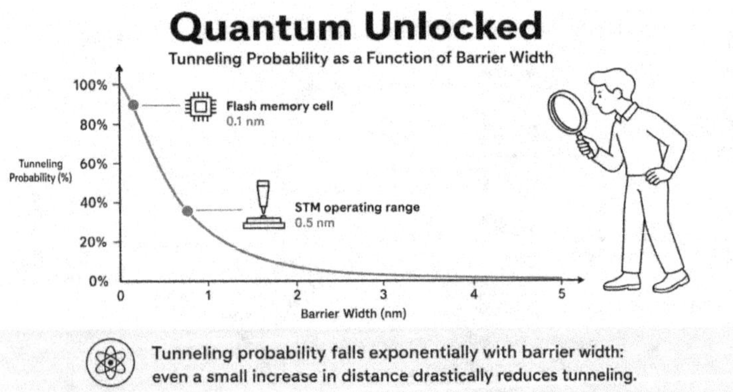


## Alpha Decay: Radioactive Escape Through a Quantum Wall

One of the first places physicists recognized tunneling in nature is in a process you have almost certainly heard of: radioactive decay. Specifically, a type called alpha decay, in which an unstable nucleus ejects a package of two protons and two neutrons, called an alpha particle. This ejection is the mechanism underlying the radioactivity of elements such as uranium, thorium, and radium. And for decades before quantum mechanics arrived, it presented a genuine mystery.

Here is the puzzle. An alpha particle inside a nucleus is held by the strong nuclear force, which acts like an extremely powerful glue at short distances. But outside the nucleus, electromagnetic repulsion from the remaining protons would push it away. Between inside and outside, the forces compete, and the alpha particle faces a potential energy barrier that classical calculations showed was much higher than the particle's total energy. The particle should, by classical reasoning, never escape. Yet alpha decay happens. Uranium and radium shed alpha particles. The evidence was impossible to ignore.

## Gamow and the Quantum Solution

In 1928, George Gamow solved this puzzle by applying quantum mechanics. His insight was that the alpha particle need not go over the barrier; it can tunnel through it. Treated as a quantum object, the alpha particle has a wave function whose amplitude penetrates the barrier and reaches the outside. The tunneling probability per attempt is tiny, but the alpha particle rattles around inside the nucleus many times per second, each time making a fresh attempt. Given enough time, the accumulated probability adds up, and the particle escapes.

Gamow's calculation predicted the observed half-lives of radioactive nuclei with striking accuracy. A nucleus with a slightly different energy configuration has a dramatically different half-life because the tunneling probability is exponentially sensitive to the barrier parameters. The half-life of uranium-238 is about 4.5 billion years. The half-life of polonium-214 is about 164 microseconds. That range spanning 24 orders of magnitude in decay rate is explained by the exponential dependence of tunneling on barrier dimensions. Gamow's formula captures this enormous spread from a single tunneling calculation.

Diagram 8.6 – Alpha Decay as Quantum Tunneling Through the Nuclear Barrier

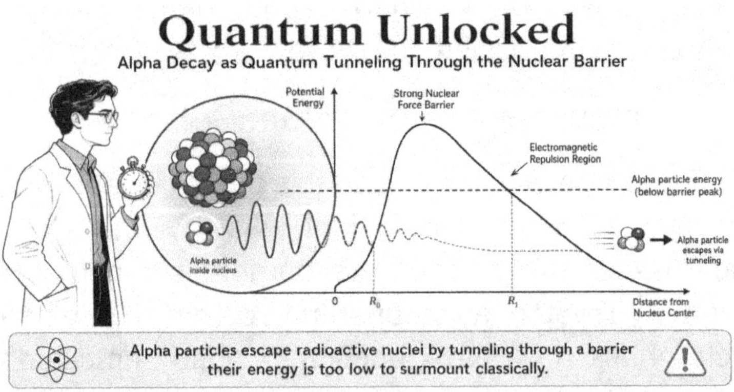


The implications extend beyond nuclear physics. The half-life data for radioactive isotopes form a massive database of tunneling events, measured with extraordinary precision. The agreement between Gamow's model and that data is one of the early triumphs of quantum mechanics. The detective

examining this evidence has no choice but to accept the verdict: particles can and do pass through barriers they classically cannot surmount.

Alpha decay also illustrates a feature of tunneling that will recur: the enormous sensitivity of the probability to small changes in parameters. A tiny increase in the alpha particle's energy above the barrier height dramatically increases the tunneling rate, thereby shortening the half-life. This is not a linear effect. Because the probability depends on an exponential, small changes in energy translate into enormous changes in decay rate. Nature exploits this sensitivity throughout nuclear physics and, as we will see, in stellar physics.

## The Engine of the Sun: Tunneling and Nuclear Fusion

If alpha decay is tunneling working in one direction, nuclear fusion in stars is tunneling working in the opposite direction. In fusion, two light nuclei approach each other and combine into a heavier nucleus, releasing energy in the process. This is the reaction that powers the sun, producing the light and heat that make life on Earth possible. And it faces the same kind of barrier problem that alpha decay does, but in reverse.

Two protons approaching each other are both positively charged and repel each other. To get close enough for the strong nuclear force to take over and fuse them, they must overcome this electromagnetic barrier. Classical calculations give the required temperature in billions of degrees Celsius. The core of the sun is hot, but only about 15 million degrees Celsius, far short of the classical requirement. According to classical physics, the sun should not be fusing protons and should not be shining.

## How Tunneling Lights the Sun

The resolution is tunneling. The protons do not have enough energy to go over the Coulomb barrier. They need enough energy to get close enough for their wave functions to overlap in the fusion region. At stellar core temperatures, a small fraction of protons have enough energy that tunneling through the remaining barrier is probable enough, over billions of years, to produce the observed luminosity of the sun.

No individual proton fusion event is likely at any given moment. The sun's luminosity comes from the sheer number of protons available (around 10^57) and the fact that each one has countless opportunities to attempt tunneling. The small but nonzero probability, multiplied by an astronomically large number of attempts, produces the steady energy output we observe. The sun is, in a deep sense, a quantum machine.

Diagram 8.7 – Nuclear Fusion in the Sun: Tunneling Through the Coulomb Barrier

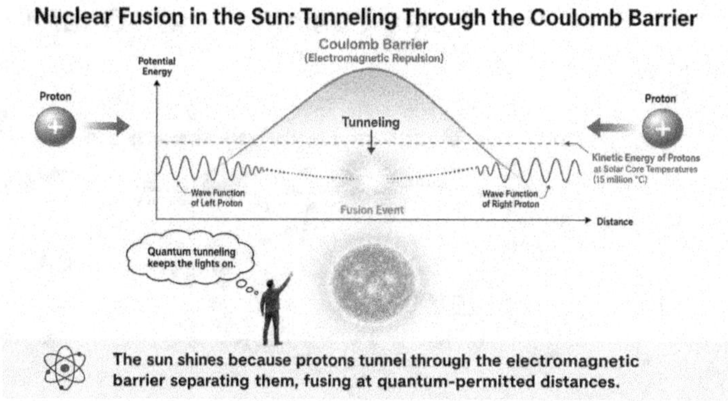


Tunneling's role in stellar fusion also explains why stars burn slowly and live long. The tunneling probability is small enough that the sun converts only a tiny fraction of its hydrogen to helium each second. If the probability were larger, the sun would exhaust its fuel far sooner. The particular value, determined by the proton mass and the dimensions of the Coulomb barrier, is one reason the sun has burned steadily for nearly five billion years and will continue to do so for another five billion. Life-bearing planets need time to evolve life. Quantum tunneling provides that time.

This is an example of what physicists call fine-tuning: the constants of nature happen to have values that allow stars to be long-lived. Whether this is a coincidence or something deeper is a question physics alone cannot answer. What is clear is that without the tunneling probability dictated by the proton mass and the electromagnetic force, the story of the universe would be dramatically different.

Diagram 8.8 – The Solar Fusion Cycle: Quantum Tunneling at Each Step

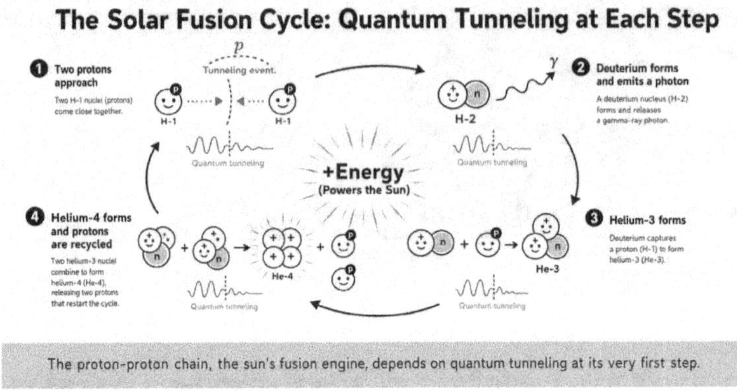


## Seeing Atoms: The Scanning Tunneling Microscope

In 1981, Gerd Binnig and Heinrich Rohrer at IBM's Zurich laboratory built a device that could do something previously thought impossible: produce images of individual atoms on a surface. For this work, they received the Nobel Prize in Physics in 1986. The device was the scanning tunneling microscope, and its operating principle is pure quantum tunneling.

An extremely sharp metal tip, ideally terminated by a single atom, is brought to within a nanometer or so of the surface. The tip and surface do not touch; a vacuum gap separates them. Classically, no current should flow across that gap. But electrons from the tip have wave functions that extend across the vacuum and overlap with the surface. Apply a small voltage, and electrons tunnel across, producing a measurable electric current.

## Using Exponential Sensitivity as a Tool

The tunneling current depends exponentially on the gap width. Reduce the gap by one-tenth of a nanometer, roughly the radius of a hydrogen atom, and the current approximately doubles. Increase it by the same amount as the current halves. This extreme sensitivity means that measuring the current as the tip scans across the surface allows a computer to reconstruct the surface height at each point with subatomic precision.

When the tip passes over a surface atom, the gap narrows and the current spikes. Over a valley between atoms, the current drops. By recording the current as a function of tip position, or by adjusting the tip height to hold the current constant and recording the resulting adjustments, the instrument maps the surface at atomic resolution. The resulting images show individual atoms arranged into patterns predicted by chemistry and crystallography.

Diagram 8.9 – The Scanning Tunneling Microscope: Tunneling Current Images Atoms

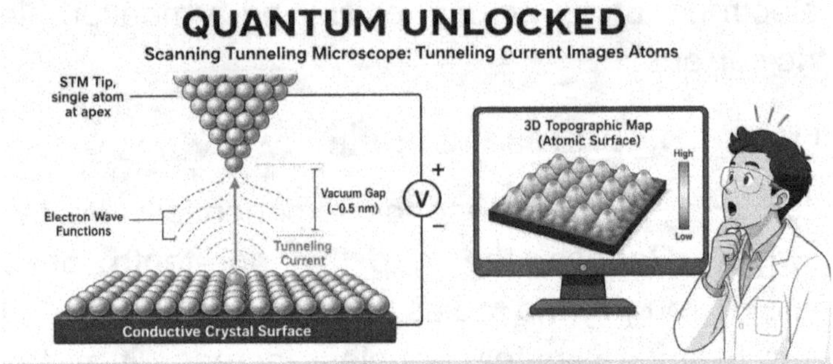


The scanning tunneling microscope images individual atoms by measuring the exponentially distance-sensitive flow of tunneling electrons.

The scanning tunneling microscope makes quantum tunneling inescapably concrete. An STM image of a silicon surface showing individual atoms in their hexagonal pattern is not an artistic impression. It is the direct consequence of quantum tunneling when it is used as a measurement tool, and its existence is proof that tunneling is real and controllable.

Beyond imaging, STM tips can be used to manipulate individual atoms, pushing them across surfaces and arranging them into patterns. In 1989, IBM researchers used an STM to spell out the letters 'IBM' by positioning 35 xenon atoms on a nickel surface. Every step of that process depended on quantum tunneling for both the detection and the manipulation. The microscope that reads atomic surfaces by exploiting tunneling was also capable of writing on them, one atom at a time.

# Tunneling at Work: Tunnel Diodes and Flash Memory

The scanning tunneling microscope is a triumph of precision science. But quantum tunneling also lives inside the devices billions of people use every day without thinking about it. Two technologies stand out for the clarity with which they exploit tunneling: tunnel diodes and flash memory.

A tunnel diode is built with a barrier so thin, typically a nanometer or less, that electrons tunnel through before they have any chance to be stopped. The result is a device with unusual electrical characteristics, including regions where increasing the voltage decreases the current, a behavior called negative differential resistance that has no classical analog.

## Flash Memory: Writing Bits with Tunneling Electrons

Flash memory, the technology used in solid-state drives and smartphones, employs a different tunneling mechanism. Each memory cell contains a floating gate, a conductor surrounded by a thin insulating layer. To store a bit, electrons are injected into the floating gate by a voltage high enough to force tunneling through the insulator. Once inside, they are trapped: the tunneling probability back out is negligible under normal conditions. To erase the bit, a reversed voltage forces the electrons to tunnel back out.

This process of writing and erasing with tunneling electrons is called Fowler-Nordheim tunneling, after the physicists who first analyzed it. Every time you save a file to a USB drive or update a photo on your phone, trillions of electrons are tunneling through insulating barriers inside the memory chips. The information in digital form is encoded in whether particular floating gates are charged or not, and that charging and discharging are accomplished purely by quantum tunneling.

Diagram 8.10 – Flash Memory Cell: Fowler-Nordheim Tunneling Writes and Erases Data

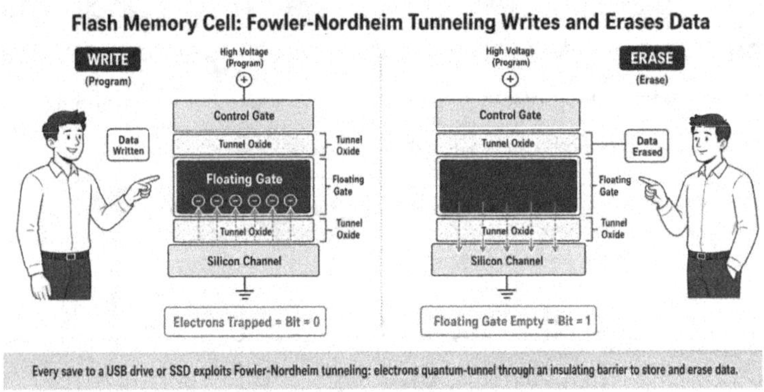


The reliability of flash memory cells is, in a sense, an engineering problem in tunneling management. A cell can be written and erased only a finite number of times before the insulating oxide degrades, because each tunneling event subtly damages the material. The 'endurance' rating for flash memory, typically in the tens of thousands of write-erase cycles, is essentially a tunneling-damage budget.

Tunnel diodes found their niche in high-frequency electronics and oscillator circuits. They are less ubiquitous than flash memory, but they were historically important as the first semiconductor device whose operation required an explicitly quantum-mechanical explanation, a fact their inventor, Leo Esaki, acknowledged when he received the Nobel Prize in Physics in 1973.

## Looking Deeper: What Tunneling Tells Us About the Quantum World

Tunneling is not just a practical phenomenon. It is a window into what quantum mechanics is saying about the nature of particles and reality. Each application raises questions that classical physics cannot even formulate.

Consider the measurement problem applied to tunneling. When an alpha particle tunnels out of a nucleus, at what moment does it cease to be inside and begin to be outside? Classical mechanics says there is a definite moment. Quantum mechanics says no such sharp moment exists. Before detection, the alpha particle's wave function has amplitude both inside and outside simultaneously. Only when we detect it, by observing its ionizing track or noting the changed nucleus, does the question of inside or outside receive a definite answer.

## Tunneling Time and Its Controversies

How long does tunneling take? This question has occupied physicists for decades and remains, in a technical sense, unresolved. Several analyses suggest the effective transit time through a barrier is surprisingly short, possibly zero in some formal treatments. Experiments have produced results suggesting that photons can tunnel through barriers faster than light would travel the same distance in a vacuum, sometimes described as 'faster than light.'

The careful interpretation is more subtle. These experiments do not violate relativity. No information travels faster than light. The apparent superluminal transit times reflect subtle features of how we define the arrival time of a wave packet. The phenomenon is real, but it does not permit faster-than-light communication. It does illustrate that quantum tunneling introduces genuinely novel concepts of time and causality that classical mechanics has no vocabulary to handle.

Diagram 8.11 – Tunneling Time: Where Classical Physics Runs Out of Words

How long does tunneling take? Even in modern quantum mechanics, the answer depends on how you define the question.

What tunneling most clearly demonstrates is the inadequacy of the particle picture. A classical particle either gets through the barrier or it does not, with a definite position at every moment. Tunneling requires the wave picture: amplitude that spreads, decays, and reforms on the far side. Every STM image, every measured alpha-decay half-life, every flash-memory write cycle is a vote for the wave picture over the particle picture.

This does not mean the particle picture is useless. When we detect an alpha particle, it arrives at a specific point, ionizing a specific atom, producing a single localized track. The tension between the spread-out wave and the localized detection is the measurement problem, the deep unresolved question at the heart of quantum mechanics that the final chapter will address. For now, tunneling serves as another reminder that the quantum world operates by precise, consistent rules that are completely alien to everyday intuition.

## Takeaway

Quantum tunneling is one of the most direct demonstrations that the quantum world is not a refinement of classical mechanics but a replacement of it. The classical verdict on energy barriers is clear: if you lack the energy to go over, you stay on this side. The quantum verdict is different. Particles are waves, and waves do not stop at boundaries. They penetrate, decay within forbidden regions, and, when the barrier is thin enough, emerge on the other side. The probability of this process depends on three things: barrier height, barrier width, and particle mass. All three operate through the exponential decay of the wave function inside the barrier, making tunneling extraordinarily sensitive to small changes in geometry and energy.

This sensitivity has consequences at every scale of nature. In atomic nuclei, it explains why radioactive elements decay at the rates they do, answering a mystery that baffled physicists for two decades before Gamow's tunneling calculation in 1928. In stellar cores, it explains how the Sun produces energy despite the classical requirement for temperatures far beyond those reached in stellar interiors. In laboratories, it underpins the scanning tunneling microscope, the instrument that gave humanity its first direct images of individual atoms. In consumer electronics, it is the working principle of flash memory, the technology that stores data on nearly every digital device in use today.

Tunneling is also a lens through which to view what quantum mechanics says about reality. Particles do not have definite positions between measurements. They exist as wave functions, smeared across space, with amplitudes that respond to potentials in ways particles cannot. The forbidden region applies only to classical particles. For quantum particles, it is a region of exponential decay, and if the decay does not reach zero before the far edge, the particle has a chance of appearing on the other side. That chance is real, measurable, and consequential. The universe runs on it.

The next chapter will bring these quantum phenomena into direct contact with the technology that defines modern civilization. Transistors, lasers, MRI machines, and the emerging world of quantum computing: all of these devices are, at their core, quantum machines. What you have learned about wave functions, superposition, entanglement, and tunneling is not preparatory material for understanding real technology. It is the real technology, translated into the language of everyday experience.

# 9 The Quantum World at Work: From Lasers to Quantum Computers

Imagine you are holding a small rectangle of glass and metal. It fits in your palm. It plays music, routes phone calls, translates speech from one language to another, and takes photographs sharper than a film camera from twenty years ago. Inside it, there are several billion transistors, each smaller than a virus. There is a laser etching data onto memory. There is a tiny antenna whose timing depends on the quantum energy levels of cesium atoms. And somewhere on its chip, engineers are beginning to experiment with circuits that do not process information in the old way at all.

You are not holding a piece of classical physics. You are holding a quantum machine.

This chapter is the payoff. The quantized energy levels, the wave nature of electrons, the spin of particles, and the collapse of superpositions are not just strange philosophy — they are working engineering. The transistor exists because quantum mechanics governs which electrons can and cannot move through a crystal. The laser exists because Einstein predicted in 1917 that a photon could coax another atom into releasing an identical photon. The MRI scanner exists because protons spin like tiny tops, and their spin states can be nudged, read, and used to create images of soft tissue. Your role as investigator is to trace the path from that abstract physics to the concrete machines of everyday life.

## Why It Matters

Every major technology of the digital age is, at its foundation, a quantum machine. Classical physics -- the physics of Newton and Maxwell -- cannot explain why silicon conducts electricity under some conditions and refuses under others. It cannot explain how a laser produces perfectly coordinated light across billions of photons. It cannot explain why hydrogen nuclei in a magnetic field absorb radio waves at a precise frequency that varies with the molecule they are in.

Quantum mechanics explains all of these things. The band structure of semiconductors, the mechanism of stimulated emission, the Larmor precession of nuclear spins -- these are quantum phenomena. Without them, the devices would be impossible. Without the devices, the modern world would be unrecognizable.

And then, on the frontier of this story, there is a new category of machine: the quantum computer. It does not merely use quantum phenomena as a side effect of its materials. It uses superposition and entanglement as the actual computational medium. Understanding what that means—and what it does not—will be one of the most important scientific literacy tasks of the coming decade.

# Semiconductors and the Band Gap: How Quantum Mechanics Built the Digital Age

## The Problem with Classical Conductors

When physicists in the early twentieth century tried to explain electrical conduction using classical physics, they ran into a wall. The Drude model -- which treated electrons in a metal as a gas of charged billiard balls bouncing through a lattice of ions -- gave roughly the right answer for some metals but could not explain why some materials conducted electricity and others did not, and why a handful of materials conducted under some conditions but not others.

The answer required quantum mechanics. Specifically, it required understanding what happens to electron energy levels when you take not one atom but 10^18 atoms and pack them into a crystal.

## Energy Bands: What Happens When Atoms Crowd Together

In an isolated atom, electrons occupy discrete, quantized energy levels. These are the specific rungs on the energy ladder discussed in Chapter 3. An electron can be on one rung or another, but never between rungs.

Now imagine bringing a second atom close to the first. The electron wave functions of the two atoms begin to overlap. The Pauli exclusion principle -- the rule that no two fermions can occupy the same quantum state -- forces the energy levels to split. Two atoms close together produce two slightly different energy levels where there was one before. Bring in a third atom, and there are three slightly different levels. Bring in ten billion billion atoms, and those discrete levels smear into a continuous band -- a range of allowed energies so densely packed that they are practically continuous.

The result is two key bands. The valence band is the range of energies occupied by the outermost electrons at low temperature. Above it, separated by a gap of forbidden energies, is the conduction band -- the range of energies an electron would need to reach to flow freely through the material.

The energy gap between these bands is called the band gap. It is the single most important quantum property for understanding why materials conduct, insulate, or behave as semiconductors.

Diagram 9.1 -- Semiconductor Band Gap: Valence Band, Conduction Band, and the Forbidden Gap

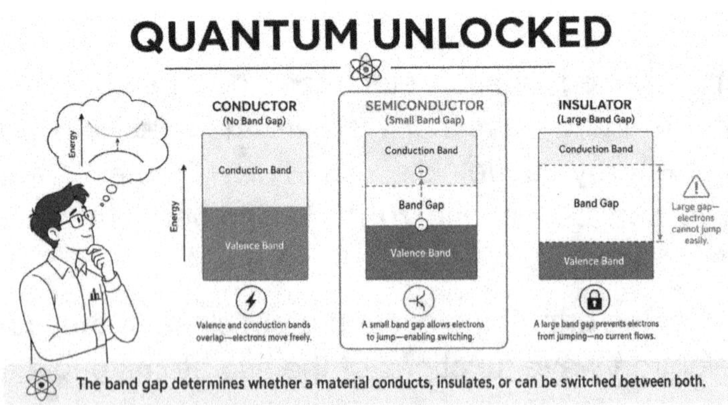

In a conductor such as copper, the valence and conduction bands overlap. Electrons can flow freely at any temperature. In an insulator such as glass or rubber, the band gap is so large -- typically several electron-volts -- that no ordinary thermal energy can push an electron across it. In a semiconductor such as silicon or germanium, the band gap is small: about 1.1 electron-volts for silicon. That is small enough that thermal energy at room temperature kicks a small fraction of electrons into the conduction band, making silicon a modest conductor. But the conductivity can be tuned.

## Doping: Engineering the Band Gap

The genius of semiconductor technology is that you can tune conductivity by introducing tiny impurity atoms -- a process called doping. Add phosphorus atoms -- which have one extra electron -- into silicon, and those extra electrons sit near the conduction band, needing only a tiny nudge to become free charge carriers: n-type doping. Add boron, which has one fewer electron, and you create positively charged holes that move through the crystal like bubbles: p-type doping.

Press a layer of n-type silicon against p-type silicon, and you create a p-n junction. A thin depletion region forms with an internal electric field. Apply voltage in one direction, and the barrier lowers: current flows. Reverse the voltage, and the barrier rises: current stops. You have built a diode, a one-way valve for electric current. The transistor is one step further.

## The Transistor: A Quantum Switch

A transistor is essentially two p-n junctions back-to-back, with a thin control gate in the middle. A small gate voltage opens or closes a channel that allows current to flow between the outer terminals. This is a switch built entirely on the quantum-mechanical band structure of silicon. Without the Pauli exclusion principle and quantized energy levels, the band gap does not exist, and the transistor does not work. Classical physics cannot explain it. The transistor is a quantum device.

Diagram 9.2 -- The p-n Junction and Transistor Operation

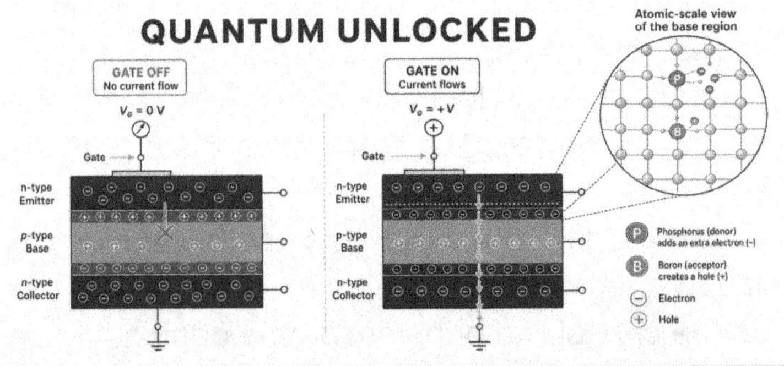

A small gate voltage controls a large current: the transistor as a quantum-mechanical switch.

Modern transistors have gate lengths of around 3 nanometers—roughly 10 silicon atoms across. At that scale, electrons tunnel through barriers that should stop them, and wave-function effects become dominant engineering challenges. Every smartphone chip contains somewhere between 10 billion and 20 billion of these quantum switches, each toggling billions of times per second.

## Stimulated Emission and Lasers: How Einstein's Prediction Became a Tool

### Three Ways an Atom Can Interact with Light

In 1917, Albert Einstein was not building a laser. He was trying to derive Planck's blackbody radiation formula from first principles, using quantum mechanics. In doing so, he realized that there had to be three distinct processes by which atoms interact with photons.

The first is absorption: an atom in a low-energy state absorbs a photon and transitions to a higher-energy state. The photon disappears. The atom is now excited.

The second is spontaneous emission: an atom in an excited state, left to itself, eventually drops back to a lower energy state and releases a photon. The photon goes in a random direction and has a random phase. This is how a light bulb and a candle produce light -- a chaotic spray of photons with no coordination.

The third process is the one Einstein discovered he had to include to make his calculation work. He called it stimulated emission. If an excited atom is hit by a photon whose energy exactly matches the energy difference between the excited state and the lower state, the incoming photon does not get absorbed. Instead, it triggers the atom to emit a second photon -- and the emitted photon is identical to the incoming one. Same energy, same direction, same phase, same polarization. The two photons are perfectly coordinated.

Einstein included this third term because the math demanded it. He had no idea, in 1917, that anyone would ever build a device to exploit it. It took until 1960 for Theodore Maiman to fire the first working laser, using a synthetic ruby rod and a flashlamp.

## How a Laser Works, Step by Step

The word laser is an acronym for "Light Amplification by Stimulated Emission of Radiation." Inside a laser cavity, a gain medium provides atoms with the appropriate energy levels — a ruby crystal in Maiman's first laser, a gas mixture, a semiconductor diode, or a doped glass fiber in modern ones. An energy pump (a flashlamp, electric current, or another laser) kicks atoms into an excited state.

The pump's goal is to achieve a population inversion: more atoms in the excited state than in the ground state. In ordinary thermal equilibrium, lower states are more populated, so photons are more likely to be absorbed than to trigger stimulated emission. Flip the population, and a photon traveling through the medium is more likely to trigger an emission event, spawning an identical twin. The medium amplifies light rather than absorbing it.

Diagram 9.3 -- Laser Operation: Population Inversion and Stimulated Emission Chain Reaction

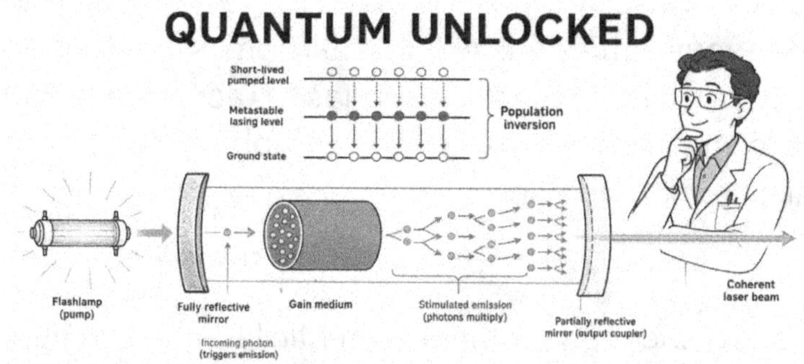


Population inversion turns the gain medium into an amplifier: each photon summons an identical twin.

With the population inversion achieved, place the gain medium between two mirrors—one fully reflective, one partially reflective. A photon emitted by spontaneous emission travels along the cavity axis, reflects off the fully reflective mirror, and passes through the gain medium again, triggering further stimulated emission. The photon count grows exponentially with each pass. The partially reflective mirror lets a small fraction of photons escape, and that escaping fraction is the laser beam.

The result is light with three remarkable properties. First, it is coherent: all the photons are in phase with each other, with their wave crests and troughs perfectly aligned. Second, it is monochromatic: all the photons have the same wavelength, determined by the energy difference between the two relevant levels in the gain medium. Third, it is collimated: all the photons travel in the same direction, with almost no divergence over long distances.

None of these properties can be achieved with spontaneous emission. A light bulb emits photons in random directions, at random phases, across a broad range of frequencies. A laser emits photons that are identical to one another.

## Lasers in Everyday Life

The coherence and monochromaticity of laser light explain why lasers are useful across such a wide range of applications. A barcode scanner uses a low-power diode laser whose perfectly collimated beam reflects from dark and light stripes at a precise angle. A Blu-ray disc uses a laser with a wavelength of 405 nanometers (blue-violet light) because the shorter wavelength allows a smaller focused spot, enabling higher data density.

Fiber optic cables carry internet traffic as pulses of laser light near 1550 nanometers, traveling through glass with very low loss. Surgical lasers cut tissue with precision no blade can match. The gravitational wave detectors at LIGO use laser light to detect distortions in spacetime smaller than one ten-thousandth the diameter of a proton. All of these trace back to Einstein's 1917 calculation that an incoming photon can trigger an identical outgoing one.

Diagram 9.4 -- Lasers Across Applications: From Data Storage to Surgery to Space

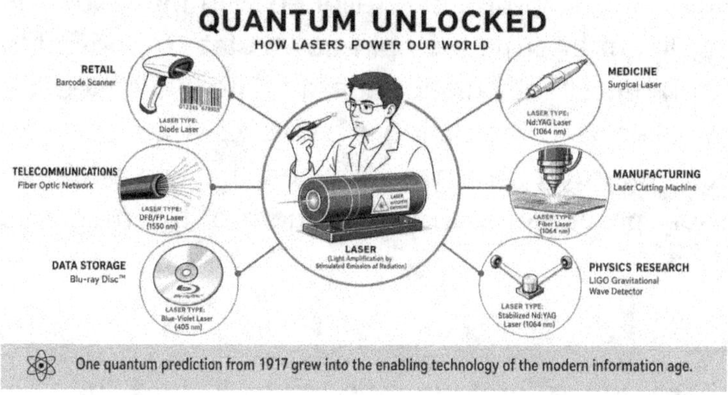


# Nuclear Magnetic Resonance and MRI: Reading Quantum Spin

## The Quantum Spin of a Proton

Every proton has an intrinsic quantum mechanical property called spin. As discussed in Chapter 5, spin is not the proton's physical rotation. It is a fundamental quantum property that gives the proton a small magnetic moment -- a tiny bar magnet built into the particle itself.

A proton's spin can point in only two directions along any measurement axis: up or down. In quantum language, its spin quantum number is one-half, giving two possible states: spin-up and spin-down. In the absence of a magnetic field, these two states have the same energy. There is no preference between them.

Apply an external magnetic field, and something changes. The proton's magnetic moment interacts with the field. The spin-up state (aligned with the field) has slightly lower energy; the spin-down state (opposed to the field) has slightly higher energy. The two states are split in energy. This splitting is called Zeeman splitting, and the energy difference between the two states is precisely proportional to the strength of the magnetic field.

## Resonance: Matching the Photon to the Gap

Bombard those protons with radio waves. At the precise Larmor frequency -- the frequency where photon energy exactly matches the Zeeman gap -- protons absorb the photon and flip spin state. This is nuclear magnetic resonance (NMR). When the pulse ends, flipped protons return to their lower-energy state, re-emitting radio photons as they do. A receiver coil captures that return signal.

Diagram 9.5 -- Nuclear Magnetic Resonance: Proton Spin Flipping in a Magnetic Field

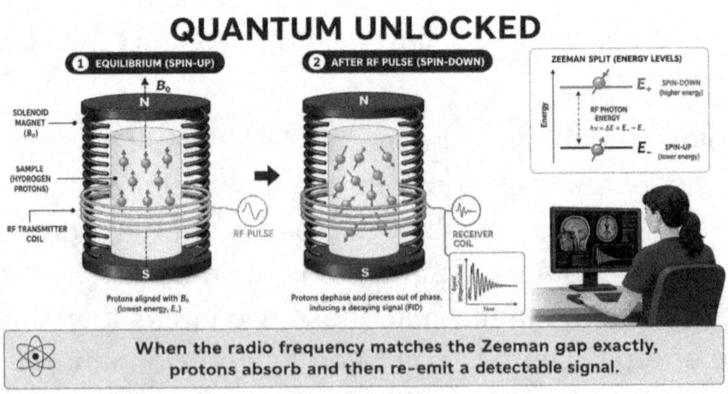


When the radio frequency matches the Zeeman gap exactly, protons absorb and then re-emit a detectable signal.

The Larmor frequency depends not only on the external field but also on the proton's local chemical environment. A proton in water resonates at a slightly different frequency than one in fat or protein. This chemical shift is the basis of NMR spectroscopy. It is also the key to imaging.

## From Bench Chemistry to Whole-Body Imaging

In the early 1970s, Paul Lauterbur realized that if you apply a spatially varying magnetic field — a field gradient — across a sample, each proton experiences a slightly different field strength depending on its position. Each position, therefore, has a slightly different Larmor frequency. By measuring which frequencies are absorbed and re-emitted, you can reconstruct where in the sample those protons are. You can make an image.

This is the principle of magnetic resonance imaging. The scanner surrounds the patient with a powerful superconducting magnet, typically producing a field of 1.5 to 3 Tesla, tens of thousands of times stronger than Earth's magnetic field. Radio-frequency pulses flip the proton spins within a thin slice of the body. Gradient coils then impose position-dependent field variations. The return signal is measured, Fourier-transformed, and reconstructed into an image.

The result is a map of proton density and relaxation rates in the body. Soft tissue -- brain, muscle, fat, tumors -- shows up with remarkable contrast, precisely because the quantum mechanical environment of each type of tissue differs. Bone, which has few free protons, appears dark. Cerebrospinal fluid appears bright. A tumor, whose tissue has different water content and molecular composition from healthy tissue, shows up with different contrast.

Diagram 9.6 -- How an MRI Scanner Works: From Quantum Spin to Full-Body Image

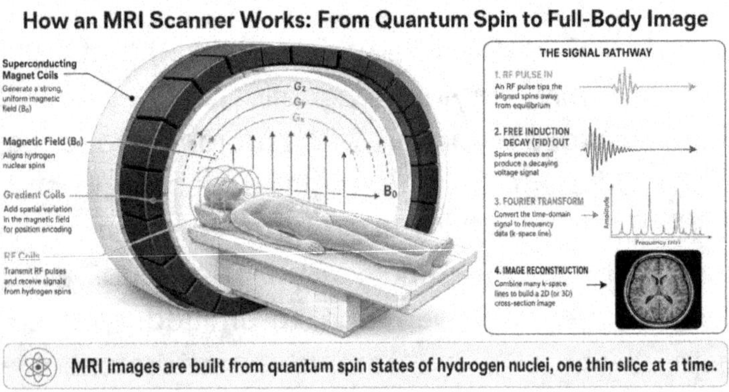


MRI is a remarkable example of quantum mechanics applied to medicine. The entire imaging chain -- from the Zeeman splitting of proton spin states to the Fourier reconstruction of a brain slice -- rests on quantum mechanical principles. Without spin, without quantized energy levels, without the precise relationship between field strength and resonance frequency, the machine does not work.

More than 40 million MRI scans are performed annually in the United States alone. Every one of them is a quantum measurement.

## Quantum Computing: Thinking in Superpositions

### The Classical Bit and Its Limitations

Every classical computer, from the first ENIAC to the latest smartphone, stores and processes information using bits. A bit is a physical system with two states: 0 and 1. In practice, a bit might be the voltage across a capacitor, the magnetization direction in a thin film, or the charge level stored in a transistor gate. Regardless of the physical implementation, it is always two and only two states. Every computation is a sequence of operations on strings of these binary values.

Classical computation is powerful, but certain problems are hard in a mathematical sense. The most infamous is factoring large numbers: the time required grows exponentially with their size. This hardness is the foundation of modern encryption. A quantum computer does not work this way. It exploits superposition, entanglement, and interference to solve certain classes of problems with dramatically fewer steps.

### The Qubit: Superposition as a Resource

The quantum bit, or qubit, is a physical system with two basis states -- typically labeled 0 and 1, corresponding to spin-up and spin-down, or two energy levels, or two polarization states of a photon. But unlike a classical bit, a qubit can exist in a superposition of both states simultaneously.

This superposition is not the same as being randomly one or the other. A qubit in superposition is in both states at once, in a mathematically precise sense. The state is described by two complex numbers -- the probability amplitudes for each outcome -- and these amplitudes can interfere with each other, just as the probability amplitudes of a particle in the double-slit experiment interfered.

Two qubits in superposition can represent all four combinations—00, 01, 10, 11—simultaneously. In general, n qubits represent 2-to-the-n states simultaneously. For 300 qubits, that number exceeds the count of atoms in the observable universe.

Diagram 9.7 -- The Qubit vs. the Classical Bit: Superposition as Parallelism

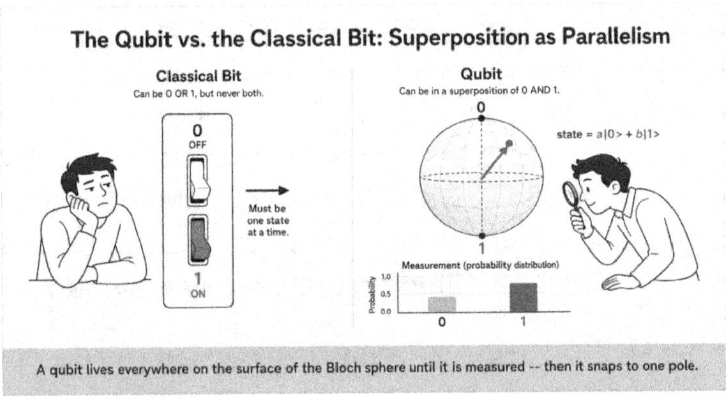


A quantum computation operates on all those states simultaneously through the unitary evolution of the

wave function. Quantum gates rotate and entangle qubits; interference between probability amplitudes suppresses wrong answers and reinforces right ones. At the end, measurement collapses the result to the correct answer with high probability.

## Entanglement as a Computational Resource

Entanglement is an essential resource, not a philosophical curiosity. When two qubits are entangled, the state of one is correlated with the other in a way that has no classical analog. Superposition creates a vast space of parallel states; entanglement weaves correlations among them, allowing interference to steer toward the correct answer. Without entanglement, no quantum algorithm outperforms its classical counterpart.

Diagram 9.8 -- Entanglement in a Quantum Circuit: Two Qubits Becoming One System

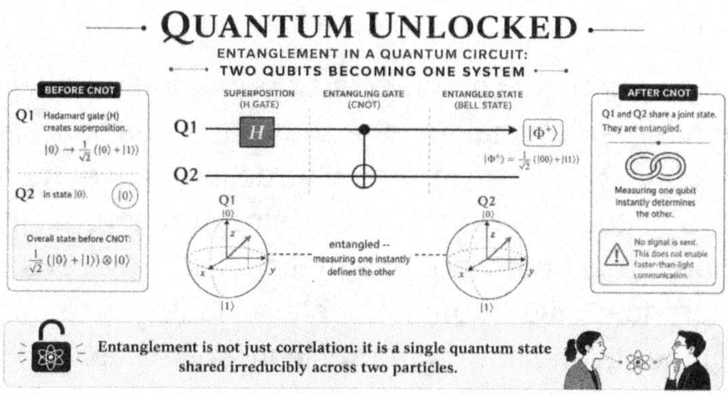

## Quantum Algorithms: What Can a Quantum Computer Actually Do?

Quantum computers are not universally faster than classical computers. For most everyday tasks, they offer no advantage. They excel at a specific set of problem types where quantum algorithms exist.

Shor's algorithm, developed in 1994, factors large integers exponentially faster than any known classical approach. A classical computer factoring a 2048-bit number would require longer than the age of the universe; a large enough quantum machine running Shor's algorithm could do it in hours. Grover's algorithm provides a quadratic speedup for unstructured search — finding one item in N takes roughly sqrt (N) steps, rather than N/2. Useful, but less dramatic than Shor's exponential advantage.

## The Hardware Challenge: Building a Qubit That Works

Building a quantum computer is extraordinarily difficult. The fundamental challenge is that superposition and entanglement are fragile. Any interaction between the qubits and the environment -- a stray photon, a tiny vibration, a fluctuating magnetic field -- causes decoherence: the quantum state leaks into the environment, and the superposition collapses. Once decoherence occurs, the computation fails.

Leading platforms each make different tradeoffs. Superconducting qubits are tiny circuits cooled near absolute zero. Trapped-ion qubits are individual ions suspended in a vacuum and manipulated by lasers. Photonic qubits encode information in photon states. Each approach has coherence times measured in microseconds to milliseconds—short compared to any meaningful computation. Quantum error correction can extend effective coherence, but requires many physical qubits per logical qubit, a large overhead that current machines cannot yet afford.

Diagram 9.9 -- Quantum Computer Hardware Platforms: Four Approaches to the Qubit

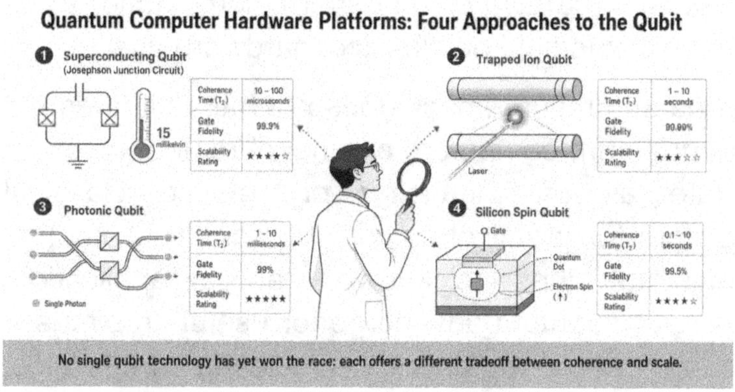


## Quantum Supremacy: What It Means and What It Does Not

In October 2019, Google announced that its Sycamore processor had achieved quantum supremacy: the first demonstration of a quantum computer completing a specific task faster than any classical supercomputer. The task was sampling from a random quantum circuit output distribution. Google reported 200 seconds on

Sycamore versus an estimated 10,000 years on Summit. IBM disputed the estimate, arguing that Summit could do it in 2.5 days with optimized algorithms. The dispute illustrated that quantum advantage is a moving target: classical algorithms improve as well.

What quantum supremacy means: a quantum processor can, in principle, execute certain quantum mechanical operations faster than any known classical approach. This is a genuine milestone. It demonstrates that quantum hardware can perform computations that would be intractable on classical hardware at scale.

What quantum supremacy does not mean: it does not mean quantum computers can now break encryption, simulate all of chemistry, or replace classical computers for practical tasks. The sampling problem demonstrated has no known practical application. Real-world quantum advantage for useful problems -- factoring large numbers, simulating molecules, optimizing logistics -- requires many more error-corrected logical qubits than any current machine possesses.

The path from today's noisy intermediate-scale quantum devices to fault-tolerant machines with thousands of logical qubits is long and technically uncertain. Practical quantum computing for cryptography-relevant problems may be decades away.

Diagram 9.10 -- Quantum Supremacy vs. Quantum Advantage: The Reality Check

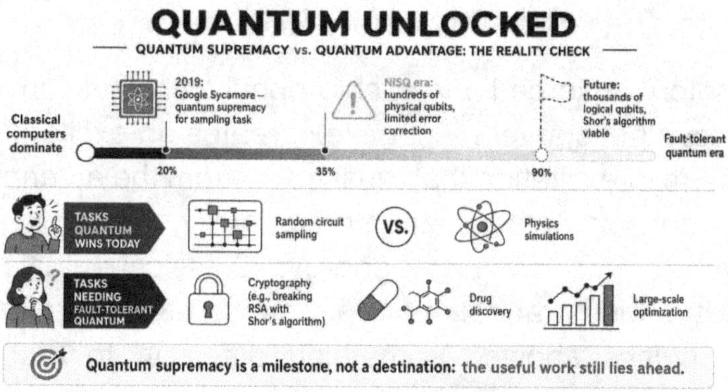


The honest picture is this: quantum computing is a genuine and important technology, not a hype cycle without substance. The physics is real. The algorithms are real. The engineering challenge is also real and formidable. The next decade will clarify which applications arrive first and which remain aspirational.

## Quantum Biology: Nature's Own Quantum Machines

### The Surprise of Quantum Effects in Living Systems

The old assumption was that living systems at body temperature are too warm and wet for quantum effects to survive. Thermal energy at room temperature dwarfs the tiny quantum gaps involved in electron tunneling or spin coherence. Quantum effects should wash out almost instantly. That assumption proved wrong in at least two striking cases: photosynthesis and the magnetic navigation of migratory birds.

## Photosynthesis: Quantum Routing of Energy

A photon absorbed by a chlorophyll molecule in a photosynthetic antenna complex creates an exciton: a quantum of excitation that must travel from the antenna to a reaction center nanometers away, where it is converted into chemical energy. Classical intuition says it diffuses randomly. Experiments using ultrafast laser pulses showed something different. In 2007, researchers found quantum coherence in the Fenna-Matthews-Olson (FMO) complex of green sulfur bacteria lasting hundreds of femtoseconds at room temperature, far longer than expected.

The exciton is not choosing a single path. It is exploring all paths simultaneously in quantum superposition. Constructive interference reinforces the route to the reaction center; destructive interference suppresses dead ends. The efficiency approaches near-perfection, far exceeding that of a classical random walk. And the protein scaffold is not merely a noisy environment: its vibrations appear to be correlated with the exciton dynamics in ways that sustain rather than destroy coherence -- a reversal of the old intuition that quantum effects require isolation.

Diagram 9.11 -- Quantum Biology: Photosynthesis and Avian Navigation

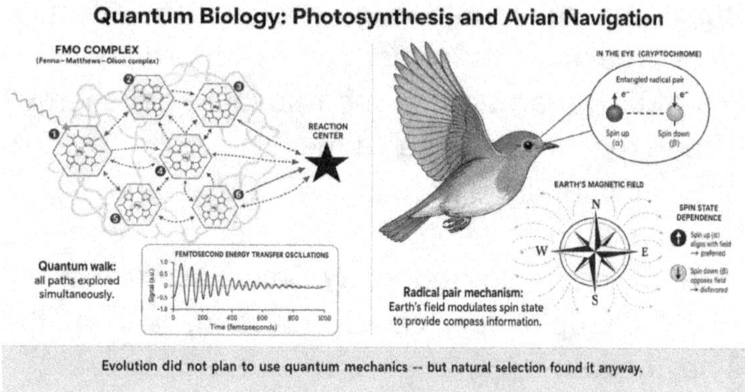


## Avian Navigation: A Quantum Compass in a Robin's Eye

European robins migrate thousands of kilometers with extraordinary precision, navigating even under overcast skies. Their magnetic compass relies not on iron-based magnetite but on quantum entanglement. When light strikes a cryptochrome molecule in the robin's retina, a chemical reaction produces two molecules, each carrying an unpaired electron. Those two electrons are entangled: their spins are correlated across a nanometer-scale separation.

Earth's magnetic field, varying in direction and inclination across the globe, subtly shifts the rate at which the radical pair's spin state evolves. The chemical outcome of the reaction depends on the spin state at completion, and that outcome affects the robin's visual signal. The bird literally sees the direction of the magnetic field. The investigator's question is sharp: how does a room-temperature bird maintain quantum entanglement long enough to affect a chemical reaction, surrounded by thermal noise millions of times larger than the relevant spin energy? The required coherence time is only a few microseconds, and the protein environment may again provide partial protection. This is not a hypothesis alone: applying radio-frequency fields at the correct frequency disrupts spin coherence and, with it, the bird's navigation.

European robins, as your investigator's dossier now shows, are not simply biological machines running on classical chemistry. They are, at least in part, quantum devices that evolution arrived at by selection pressure alone, without any physicist's guidance.

### The Open Questions in Quantum Biology

The evidence is compelling, but the field remains young and contested. What is no longer controversial is the starting premise: quantum effects can survive in warm, wet biological environments long enough to play a functional role. That overturns the old assumption and opens a new field that may eventually reach into enzyme catalysis, DNA mutation rates, and the energetics of the brain.

## Bringing It Together: The Quantum World You Live In

Retrace the chapter and notice the common thread. The transistor works because the band gap is a quantum phenomenon with no classical explanation.

The laser works because stimulated emission produces perfectly coherent light through a quantum process that Einstein found mathematically. The MRI scanner works because proton spin is quantized and its Larmor frequency encodes molecular identity.

The quantum computer takes this relationship one step further. Where the transistor and laser use quantum effects as a kind of substrate -- the physics that makes the device possible -- the quantum computer uses quantum effects as the computational medium itself. Superposition is not a side effect. It is the algorithm.

And quantum biology suggests that evolution, optimizing living systems over billions of years of selection pressure, found quantum mechanics before engineers did. Photosynthesis may be exploiting quantum coherence to route energy with near-perfect efficiency. Birds may be exploiting quantum entanglement to read Earth's magnetic field with remarkable precision. These are not designed quantum machines -- they are naturally discovered ones.

The picture that emerges is this: quantum mechanics is not merely the hidden plumbing of a few specialized instruments. It is the operating system of the physical world, running silently beneath every transistor, every photon, every molecule, and every migration.

## Takeaway

This chapter began with a simple observation: the device in your pocket is a quantum machine. It ends with something broader. Quantum mechanics is not an abstract philosophical puzzle confined to physics laboratories. It is the engineering principle behind transistors, lasers, and MRI scanners — three technologies that, together, define the modern world of computing, communication, and medicine.

The transistor exploits the quantum band structure of silicon: a consequence of the Pauli exclusion principle acting on billions of electrons packed into a crystal. The laser exploits stimulated emission: a quantum process that Einstein predicted mathematically forty years before anyone built one. The MRI scanner exploits the quantized spin of protons and their precise magnetic resonance: a phenomenon that has no classical explanation.

Quantum computing goes one step further. It does not merely use quantum effects as a substrate: it uses superposition and entanglement as the computational medium itself. Shor's algorithm and Grover's algorithm are genuinely quantum algorithms with no classical equivalent. The hardware to run them at scale remains a formidable engineering challenge, and quantum supremacy demonstrations, while genuine milestones, are not yet evidence of practical advantage for real-world problems.

Quantum biology opens the most surprising frontier. Photosynthesis in green sulfur bacteria may exploit quantum coherence to route excitation energy with near-perfect efficiency. European robins may use quantum entanglement in radical pairs to read Earth's magnetic field as a visual signal. Evolution, optimizing living systems under selection pressure, appears to have discovered quantum mechanics long before any physicist did.

The reader takeaway is both specific and sweeping. Specifically, you can now describe, at a conceptual level, how each of these technologies exploits a distinct quantum phenomenon. Broadly: the strangeness you encountered in earlier chapters -- superposition, entanglement, quantized energy, spin -- was never merely theoretical. It was always the hidden engineering beneath the world you live in.

# 10 Many Worlds and Open Questions: What Quantum Mechanics Still Cannot Tell Us

You have reached the end of a long investigation. You have watched particles tunnel through walls. You have seen light behave as both a wave and a particle depending on how you look at it. You have traced the birth of lasers, semiconductors, and MRI machines to equations that most people never encounter. You have held in mind the strange fact that two particles on opposite sides of the galaxy can be correlated in ways that no classical message could explain. You have done all of this without a single equation.

Now comes the hardest part. Not harder because the experiments are more complicated, but harder because we reach the edge of what is settled. Quantum mechanics is the most precisely tested theory in the history of science. Its predictions match experiment to more decimal places than any other framework humanity has ever constructed. And yet, physicists still argue, passionately and without resolution, about what the theory means. What is the wave function, really? Does it describe reality, or does it describe our knowledge of reality? When a measurement happens, what exactly happens? These are not fringe questions. They are asked by working physicists at the world's best universities.

This final chapter is an honest account of a field in philosophical tension. It is not a disappointment. It is, if you are willing to sit with it, one of the most exciting facts in all of science.

## Why It Matters

Most fields of science have their interpretation settled. Chemists do not argue about whether atoms are real. Biologists do not debate whether evolution describes something that actually happened. But in quantum mechanics, the interpretation, meaning the question of what the theory is actually describing about the world, remains genuinely open after nearly a century of effort. Three major schools of thought exist. They agree on every prediction. They agree on every experimental result. They disagree about what the theory means.

That disagreement matters for a simple reason. If you want to build the physics of the future, if you want to unify quantum mechanics with general relativity, you need to know what you are working with. You need to know whether the wave function is a real physical thing or a mathematical bookkeeping device. You need to know whether there is one world or many. The interpretation is not academic. It is the foundation of the next century of physics.

Diagram 10.1 – The Three Interpretations at a Glance

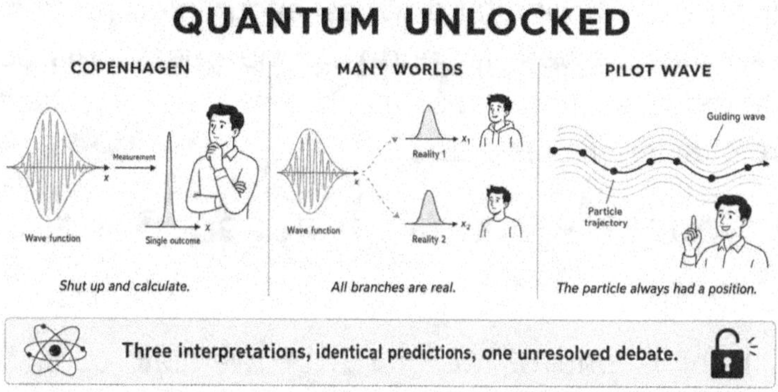


A three-column comparison diagram. The left column is labeled 'Copenhagen'. It shows a wave function collapsing into a single definite outcome upon measurement, depicted as a broad wave narrowing into a single spike, with a stylized human observer figure beside it. The center column is labeled 'Many Worlds' and shows the wave function branching into two parallel realities, each continuing as a separate thread with small human figures in both branches. The right column is labeled 'Pilot Wave' and shows a particle riding a real guiding wave, with a deterministic, continuous trajectory. Each column has a short caption below it stating the core claim: 'Shut up and calculate,' 'All branches are real,' and 'The particle always had a position.' Style: blue palette, white background, both icons and human figures, flat vector, 16:9 landscape, caption bar at bottom reading 'Three interpretations, identical predictions, one unresolved debate.'

The investigator's job in this chapter is not to pick a winner. The evidence does not yet allow that. Instead, the job is to understand what each interpretation claims, what it gives up, and why the question is so hard to resolve. That kind of intellectual honesty is itself a form of scientific progress.

## The Copenhagen Interpretation: Bohr's Pragmatic Truce

Niels Bohr is one of the giants of twentieth-century physics. His 1913 model of the hydrogen atom was the first step toward quantum mechanics. His debates with Einstein are legendary. And his interpretation of quantum mechanics, now called the Copenhagen Interpretation, was for decades the dominant view among working physicists.

The Copenhagen Interpretation is, at its core, a pragmatic refusal to speculate. Bohr argued that the wave function is a calculational tool. It allows us to predict the probability of measurement outcomes with extraordinary precision. What it does not do, and what we should not expect it to do, is describe what the particle is doing when no one is measuring it. According to Bohr, that question is not well-formed. Asking what the electron is 'really doing' between measurements is like asking what a symphony sounds like when no one is listening. The question assumes a kind of reality that quantum mechanics does not support.

## The Measurement Problem

The Copenhagen Interpretation sharpens the greatest difficulty in quantum mechanics. That difficulty is called the measurement problem. Here is the problem in plain terms.

Before a measurement, an electron in a double-slit experiment exists in a superposition of going through both slits simultaneously. The wave function encodes this superposition. When you make a measurement, the superposition collapses into a single definite outcome. The electron is either here or there, not both. Copenhagen says this collapse happens when a measurement is made. But it does not say what counts as a measurement. Is a measurement a physical interaction with a detector? Is it the registration of a result in a computer? Does it require a conscious observer?

Bohr deliberately left this boundary vague. He divided the world into the quantum domain, described by wave functions, and the classical domain of measuring devices, described by ordinary physics. The boundary between them, sometimes called the Heisenberg cut, is placed wherever it is convenient for the calculation. Critics have argued for decades that this is unsatisfying. The universe does not contain a built-in classical measuring apparatus. The apparatus is itself made of quantum particles. Drawing an arbitrary line and refusing to cross it is not a solution to the measurement problem. It is a decision to live with it.

Diagram 10.2 – The Heisenberg Cut and the Measurement Problem

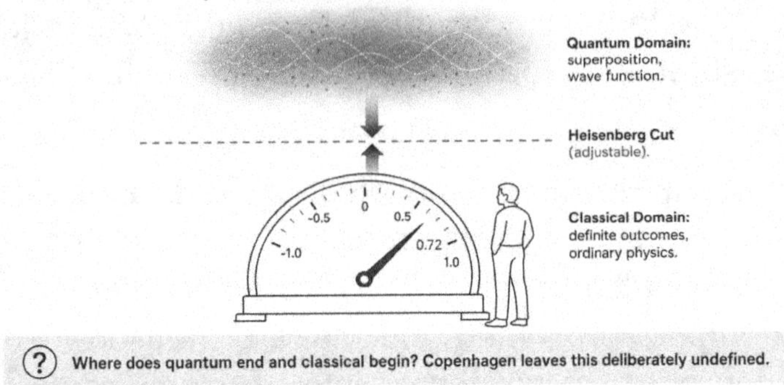


A vertical diagram showing three levels. At the top, a quantum system is depicted as a soft, blurred probability cloud labeled 'Quantum Domain: superposition, wave function.' In the middle, a dashed horizontal line is labeled 'Heisenberg Cut (adjustable).' Below the dashed line, a stylized measuring instrument with a needle pointing to a definite value is labeled 'Classical Domain: definite outcomes, ordinary physics.' A small human figure stands to the right of the instrument, reading the dial. Arrows indicate an interaction that crosses the dashed line, with the cloud above collapsing into the single pointer reading below. Style: blue palette, white background, both icons and human figures, flat vector, 16:9 landscape, caption bar at bottom reading 'Where does quantum end and classical begin? Copenhagen leaves this deliberately undefined.'

Despite its philosophical gaps, the Copenhagen Interpretation has extraordinary practical power. The generations of physicists who built quantum electrodynamics, invented the transistor, and designed the laser were, for the most part, Copenhagen practitioners. They calculated. They predicted. They built things. The philosophy could wait.

## Measurement and Consciousness

One consequence of Copenhagen's vagueness about what constitutes a measurement is a trail that leads, uncomfortably, toward consciousness. Some early interpretations, associated with figures like John von Neumann and later extended by others, suggested that the wave function collapses when it enters the awareness of a conscious observer. Not just a detector, not just a computer recording, but an actually conscious mind.

This is not the mainstream view. Most physicists who accept Copenhagen do not invoke consciousness. But the trail exists, and it is worth examining honestly. The thought experiment most often used to make it vivid is Schrodinger's cat, which you encountered in an earlier chapter. A cat is placed in a box with a quantum trigger. Until you open the box, the cat, according to a strict Copenhagen interpretation, is in a superposition of being alive and dead. Opening the box and measuring it resolves the superposition.

The question the thought experiment forces is this: what is special about a human observer? If a camera records the cat, has the wave function collapsed? If a bacterium touches the trigger mechanism, has the collapse happened? Copenhagen does not give a clean answer. This is not a criticism lodged by outsiders. It is a gap that Bohr's own colleagues, including Einstein, pressed repeatedly. It remains open.

Diagram 10.3 – Measurement, Consciousness, and the Chain of Observation

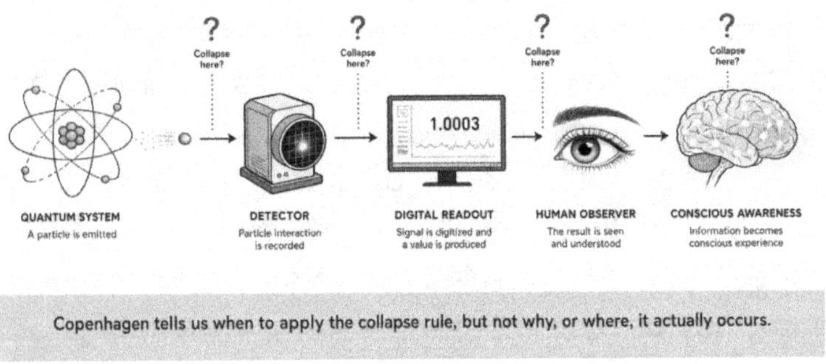


A horizontal chain-of-causation diagram. On the far left, a stylized atom emits a particle. An arrow points to a detector that registers a signal. Another arrow leads to a computer display showing a number. Another arrow points to a human eye, then to a stylized human brain with soft, glowing lines inside it. Above each link in the chain, a question mark floats: 'Collapse here? Or here? Or here?' The atom and detector are shown in blue tones; the brain is shown in slightly warmer tones to indicate the ambiguity. Style: blue palette, white background, both icons and human figures, flat vector, 16:9 landscape, caption bar at bottom reading

'Copenhagen tells us when to apply the collapse rule, but not why, or where, it actually occurs.'

## The Many Worlds Interpretation: Everett's Radical Proposal

In 1957, a graduate student at Princeton named Hugh Everett III submitted a doctoral thesis that proposed a radical solution to the measurement problem. His solution was simple in its logic and staggering in its implications: there is no collapse. The wave function never collapses. It always evolves according to the Schrodinger equation, without exception, forever.

What, then, explains the fact that you always see a definite outcome when you perform a measurement? Everett answered that when a measurement occurs, the universe branches. Every possible outcome of the measurement happens in a separate branch of reality. You, the observer, become entangled with the measurement outcome in your branch. You see a definite result, not because the other possibilities ceased to exist, but because you are now in a specific branch where one outcome occurred. The other branches, containing versions of you who saw different results, are equally real. They do not interact with your branch.

### What Branching Means

The Many Worlds Interpretation has attracted a serious and growing community of defenders. Its appeal is simplicity. It adds nothing to quantum mechanics. It does not invent a new collapse mechanism. It does not draw an arbitrary boundary between quantum and classical domains. It takes the Schrodinger equation seriously and follows it without compromise.

The cost is ontological extravagance. If every quantum event branches the universe, and quantum events are happening constantly at the level of every atom in your body, then the number of branches is incomprehensibly large. Effectively infinite. There is a version of you, in some branch of reality, reading this sentence under entirely different circumstances. Most physicists, when they first encounter this, experience something between wonder and unease.

Diagram 10.4 – Branching Universe Diagram (Many Worlds)

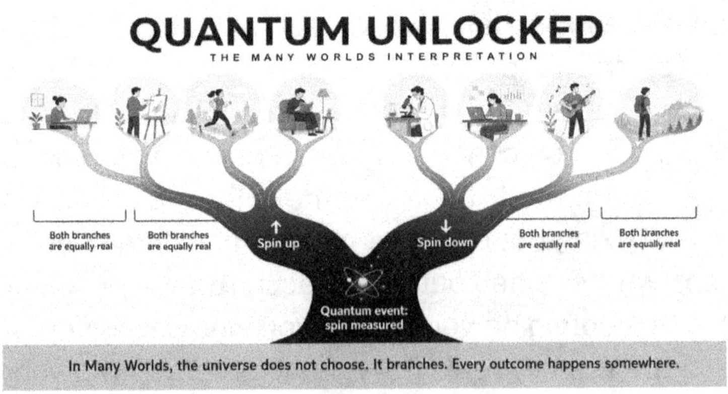


A tree-structure diagram with a single trunk at the bottom labeled 'Quantum event: spin measured.' The trunk splits into two branches, each labeled with an outcome: 'Spin up' on the left and 'Spin down' on the right. Each of those branches then splits again at a second quantum event, producing four smaller branches, each carrying small human figures going about their lives in slightly different circumstances. The branches are drawn in progressively lighter shades of blue as they extend upward, suggesting the vast proliferation of reality. A soft caption beneath each pair of branches reads, 'Both branches are equally real.' Style: blue palette, white background, both icons and human figures, flat vector, 16:9 landscape, caption bar at bottom reading 'In Many Worlds, the universe does not choose. It branches. Every outcome happens somewhere.'

The Many Worlds Interpretation faces serious challenges. The deepest problem is that of probability. In a world where every outcome happens, what does it mean to say that one outcome is more probable than another? The Born rule must somehow emerge from a theory where nothing is ever eliminated. Defenders have proposed decision-theoretic arguments, but the question remains unresolved. A second challenge is that branches do not interact, so their existence is, in principle, experimentally undetectable. Critics say this crosses from science into metaphysics. Defenders reply that the branches are implied by the same equations that make the testable predictions so accurate.

## Modern Defenders and the Oxford School

Many Worlds went through decades of neglect after Everett's original paper received a cold reception. His advisor, John Wheeler, initially supported it but later distanced himself from it. Everett left academia, and the interpretation languished. In the 1990s and 2000s, physicists and philosophers centered at Oxford, including David Deutsch and David Wallace, revived the Everett program with new mathematical rigor. They argue that decoherence provides the mechanism by which branches become effectively independent, and that the probability problem can be solved using decision-theoretic arguments about rational agents. The debate continues, but Many Worlds is no longer a fringe position.

Diagram 10.5 – The Probability Problem in Many Worlds

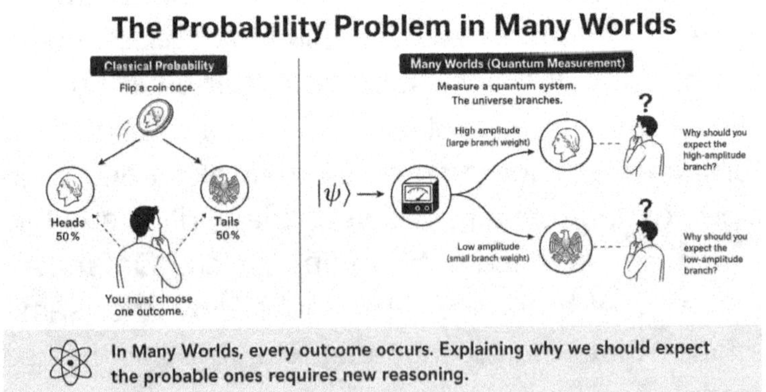


A split diagram. On the left, a classical probability scenario depicts a coin toss labeled 'Heads 50%' and 'Tails 50%', with a human figure choosing between the two outcomes. On the right side, a Many Worlds scenario shows a quantum measurement branching into an outcome with high quantum amplitude (labeled 'large branch weight') and one with low amplitude (labeled 'small branch weight'). A stylized human figure in each branch looks identical, with a question mark above each figure and the caption 'Why should you expect the high-amplitude branch?' The diagram illustrates the Born rule problem visually. Style: blue palette, white background, both icons and human figures, flat vector, 16:9 landscape, caption bar at bottom reading 'In Many Worlds, every outcome occurs. Explaining why we should expect the probable ones requires new reasoning.'

## Pilot Wave Theory: de Broglie and Bohm's Deterministic Alternative

There is a third option that many people, including many physicists, do not know exists. It is called pilot wave theory, and its history stretches back to the very beginning of quantum mechanics.

In 1927, Louis de Broglie proposed that quantum particles are not merely waves of probability. They are real particles, with definite positions at all times, guided through space by a real physical wave. The wave is real. The particle is real. The apparent randomness of quantum mechanics arises not from the universe being fundamentally random, but from our ignorance of the particle's precise starting position. If you knew the exact starting conditions, you could, in principle, predict every outcome deterministically.

## Bohm's Completion

De Broglie's original proposal was incomplete, and it was criticized sharply at the 1927 Solvay Conference, the famous gathering of the founders of quantum mechanics. De Broglie abandoned the idea. It lay dormant for twenty-five years.

In 1952, David Bohm independently rediscovered and completed the theory. Bohm was working at Princeton, having been forced out of a position at Berkeley during the McCarthy-era political purges. His reformulation was rigorous and complete. It reproduced every prediction of standard quantum mechanics exactly. It was deterministic. It was realistic, meaning particles always have definite positions. And it was invisible to criticism on grounds of empirical failure, because there was no empirical failure.

Einstein, who had spent decades insisting that quantum mechanics was incomplete, received Bohm's paper with cautious interest. He did not fully embrace it, partly because it had a feature that troubled him deeply: nonlocality. The guiding wave in Bohm's theory is defined over the entire configuration space of the system. For an entangled pair of particles, the guiding wave instantaneously connects both particles regardless of their separation. This nonlocality does not allow faster-than-light signaling, but it is a real and irreducible feature of the theory.

Diagram 10.6 – Pilot Wave Theory: Particle and Guide Wave

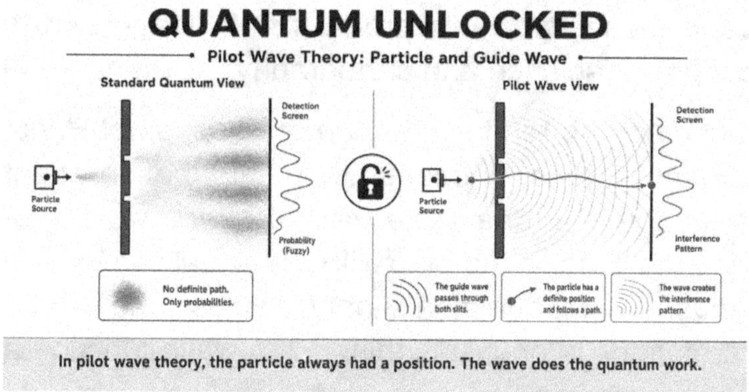


A side-by-side comparison diagram. On the left, labeled 'Standard Quantum View,' a particle is depicted as a fuzzy probability cloud spreading across a double-slit experiment, collapsing to a point upon detection. On the right, labeled 'Pilot Wave View,' a distinct particle with a definite position rides along a smooth guiding wave that passes through both slits simultaneously, creating interference that steers the particle to specific points on the detector. The particle's actual path is shown as a solid blue line curving through one slit, while the wave extends through both slits. Style: blue palette, white background, both icons and human figures, flat vector, 16:9 landscape, caption bar at bottom reading 'In pilot wave theory, the particle always had a position. The wave does the quantum work.'

## The Maps and Territories Analogy

Here is a useful way to frame the disagreement between these three interpretations. It draws on the philosophy of science and cartography.

A map is not the territory. A map of Paris is not Paris. It represents Paris, encodes useful information about Paris, and lets you navigate Paris. But you would not confuse the lines on the map with the actual streets and buildings. The question that divides the three interpretations is precisely this: is the wave function a map or the territory?

Copenhagen says the wave function is a map. It is a tool for predicting measurement outcomes. Do not ask what the territory looks like when no one is navigating. That question is not meaningful.

Many Worlds says the wave function is the territory. The wave function is literally all there is. Branching of the universe occurs when the wave function evolves according to its equation.

Pilot wave theory says both a map and a territory exist: the guiding wave is a real physical field in the territory, and the particle is a real object moving through that territory. The wave function is a complete description of the guiding wave and a map, in that it encodes everything we need to make predictions.

None of these options is absurd. Serious physicists and philosophers have defended each. The fact that three such different pictures can yield identical experimental predictions is itself one of the most remarkable features of quantum mechanics.

Diagram 10.7 – Maps vs. Territories: Three Views of the Wave Function

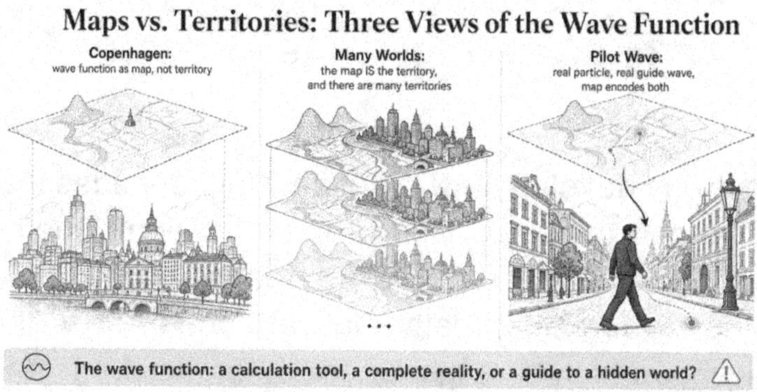


A three-panel visual metaphor diagram. The left panel shows a stylized map of a city (labeled 'Copenhagen: wave function as map, not territory') with no actual city behind it, just the map floating in white space. The center panel shows the map literally merging with the city beneath it, with parallel copies of the city branching out (labeled 'Many Worlds: the map IS the territory, and there are many territories'). The right panel shows both a map and a city co-existing, with a guiding arrow from the map pointing to a person walking a real street (labeled 'Pilot Wave: real particle, real guide wave, map encodes both'). Style: blue palette, white background, both icons and human figures, flat vector, 16:9 landscape, caption bar at bottom reading 'The wave function: a calculation tool, a complete reality, or a guide to a hidden world?'

Bell's theorem ruled out local hidden variable theories. But pilot wave theory is explicitly nonlocal. Bell himself showed that his theorem does not rule out Bohmian mechanics, and he spent part of his career arguing that it deserved more serious attention than it received.

## Decoherence and the Emergence of the Classical World

There is one development in the past four decades that has changed the landscape of interpretation debates more than any other. It is called decoherence, and it explains, in precise mathematical terms, how the classical world we experience emerges from the underlying quantum world.

The central puzzle, remember, is this: if quantum mechanics applies to everything, why do we never see cats in superpositions of alive and dead? Why does the quantum fuzziness never appear in our everyday experience? Why do objects have definite positions and definite velocities at the scale we live in?

### What Decoherence Actually Does

The answer is that quantum systems do not exist in isolation. A real particle, an electron in your body or a photon traveling across a room, is constantly interacting with its environment: other particles, electromagnetic fields, thermal radiation, and air molecules. Each of these interactions entangles the particle with its environment. As the entanglement spreads through an ever-larger number of environmental particles, the quantum coherence between different possible states of the original particle becomes effectively inaccessible. The different branches of the wave function stop interfering with each other, not because they have disappeared, but because they are now encoded in the correlations of billions of particles that cannot all be tracked and measured simultaneously.

This process is called decoherence, and it is extremely fast. For a dust grain, a particle large enough to see with a microscope but far smaller than anything visible to the naked eye, decoherence occurs in roughly 10^-36 seconds. This is so fast that it is essentially instantaneous. For an object the size of a cat, the timescale is inconceivably shorter. This is why we never see cats in superpositions. Not because the Schrodinger equation fails at large scales, but because decoherence destroys the coherence that would allow macroscopic superpositions to display quantum behavior.

Diagram 10.8 – Decoherence: From Quantum to Classical

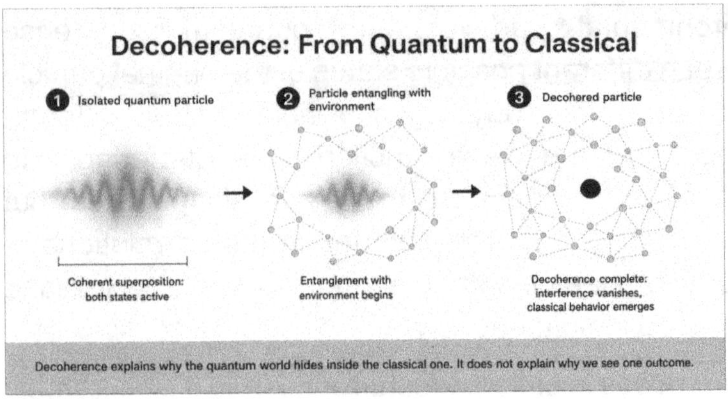


A three-stage horizontal process diagram. Stage one, on the left, shows an isolated quantum particle labeled 'Coherent superposition: both states active' as a blurred, interference-capable wave packet in blue. Stage two, in the center, shows the particle entangling with environmental particles represented as an expanding web of smaller blue dots surrounding it, labeled 'Entanglement with environment begins.' Stage three, on the right, shows the original particle now appearing classical, a solid dot with a definite position. At the same time, the environmental web surrounds it, labeled 'Decoherence complete: interference vanishes, classical behavior emerges.' Style: blue palette, white background, both icons and human figures, flat vector, 16:9 landscape, caption bar at bottom reading 'Decoherence explains why the quantum world hides inside the classical one. It does not explain why we see one outcome.'

Decoherence is not controversial. It is a precisely calculated, experimentally confirmed phenomenon.

But it does not, on its own, solve the measurement problem. Decoherence explains why different branches of the wave function stop interfering with one another. It does not explain why you, the observer, find yourself in one particular branch and not another. Copenhagen uses decoherence, but still needs the collapse rule. Many Worlds uses decoherence to define branches and argues that the question of 'which branch am I in' is answered by the structure of the branching itself. Pilot wave theory incorporates decoherence as the mechanism by which the particle's actual position becomes effectively inaccessible to macroscopic observation.

## The Preferred Basis Problem

Decoherence introduces another subtlety worth noting. When the environment measures a quantum system, there is a question of which basis, which set of possible states, the measurement occurs in. In practice, the interaction between a system and its environment selects a preferred basis: it is typically position, or states close to position, that survive decoherence most robustly. This is why the classical world is a world of definite positions and definite trajectories rather than superpositions of position. The environment acts as a continuous measurement apparatus that picks out position as the classical variable.

This selection of preferred basis by the environment is called environmentally induced superselection, or einselection, a term coined by the physicist Wojciech Zurek. The reason the world looks classical is not that quantum mechanics breaks down. It is that the environment continuously decoheres quantum systems in the position basis, rendering quantum coherence effectively invisible at macroscopic scales.

Diagram 10.9 – Decoherence and Einselection: Why Position Wins

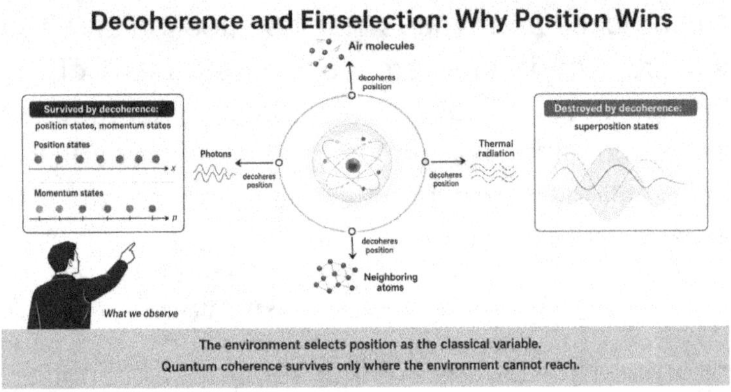


A circular diagram showing a quantum system at the center, surrounded by a ring of labeled interaction arrows pointing outward to environmental particles: 'Air molecules,' 'Photons,' 'Thermal radiation,' 'Neighboring atoms.' Each interaction arrow is labeled 'decoheres position.' Outside the ring, two boxes compare outcomes: on the left, 'Survived by decoherence: position states, momentum states' shown as clean, distinct blue dots; on the right, 'Destroyed by decoherence: superposition states' shown as faded, overlapping waves. A stylized human figure at the bottom of the diagram points to the 'Survived' box, labeled 'What we observe.' Style: blue palette, white background, both icons and human figures, flat vector, 16:9 landscape, caption bar at bottom reading 'The environment selects position as the classical variable. Quantum coherence survives only where the environment cannot reach.'

## Quantum Gravity: The Open Frontier

Set aside the debates over interpretation for a moment and consider an entirely different frontier. There is a question in physics that has gone unanswered for nearly a century, despite the efforts of the greatest minds in the field. It is the question of how to unify quantum mechanics with general relativity.

Quantum mechanics is the theory of the very small: atoms, particles, and fields at tiny scales. General relativity is the theory of the very large: gravity, spacetime curvature, and the structure of the cosmos. Both theories are extraordinarily successful in their own domains. Quantum electrodynamics predicts the electron's magnetic moment to twelve decimal places. General relativity predicted gravitational waves, which were confirmed in 2015 with instruments sensitive to displacements smaller than one thousandth of the diameter of a proton.

The problem is that the two theories are, in a precise technical sense, incompatible. They use different mathematical frameworks. They make different assumptions about the nature of space and time. And in regimes where both theories should be simultaneously relevant, such as the interior of a black hole, or the first instant of the Big Bang, the two theories give results that cannot both be correct.

## Loop Quantum Gravity

One approach to unifying quantum mechanics and gravity is called loop quantum gravity. It takes general relativity seriously as a description of spacetime and attempts to quantize spacetime itself. In loop quantum gravity, space is not a smooth, continuous medium. It is a discrete network of interconnected nodes and links, called a spin network, that exists at the Planck scale: roughly ten to the negative thirty-five meters, far smaller than anything that can currently be detected. Time, too, is not continuous. It advances in discrete steps.

Loop quantum gravity makes specific predictions about the behavior of black holes and the very early universe. One of its most striking implications is that the Big Bang may not have been a singularity, a point of infinite density, but rather a quantum bounce: a previous universe contracting to its minimum possible volume and rebounding into the one we inhabit. This prediction may, in principle, be testable through future observations of patterns in the cosmic microwave background.

Diagram 10.10 – Loop Quantum Gravity vs. String Theory: Two Roads to Quantum Gravity

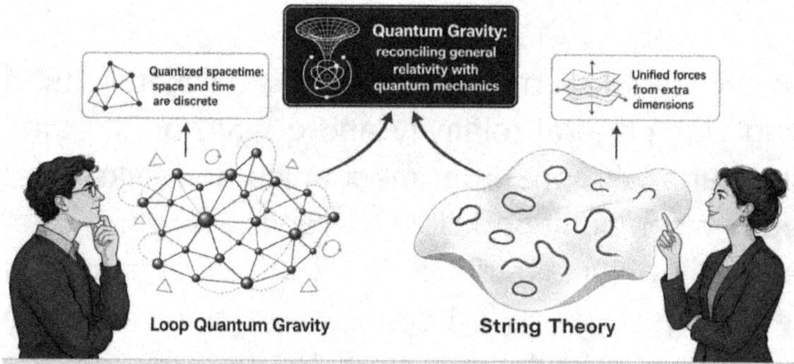


Two serious proposals, two radically different pictures of what space and time actually are at the smallest scale.

A split diagram showing two parallel routes toward a shared destination. On the left side, labeled 'Loop Quantum Gravity,' spacetime is shown as a discrete, mesh-like spin network of nodes and edges, with small geometric shapes (triangles and loops) filling the fabric of space. An arrow points upward toward a box labeled 'Quantized spacetime: space and time are discrete.' On the right side, labeled 'String Theory,' point particles are replaced by tiny vibrating strings in a smooth higher-dimensional space, with the strings shown as small loops and open segments. An arrow points upward toward a box labeled 'Unified forces from extra dimensions.' In the center, a target labeled 'Quantum Gravity: reconciling general relativity with quantum mechanics' is shown at the top, with both arrows converging toward it. Style: blue palette, white background, both icons and human figures, flat vector, 16:9 landscape, caption bar at bottom reading 'Two serious proposals, two radically different pictures of what space and time actually are at the smallest scale.'

## String Theory

String theory takes a different approach. Rather than starting with general relativity and quantizing it, string theory starts with quantum mechanics and asks what happens when point particles are replaced by tiny one-dimensional objects: strings. The vibrational modes of these strings correspond to different particles. The graviton, the hypothetical particle that carries the gravitational force, arises naturally as one of the vibrational modes.

String theory requires extra dimensions beyond the three we observe. In the most developed versions, there are ten or eleven dimensions, with the extra ones compactified at scales far too small to detect. The theory has produced remarkable mathematical insights and unexpected connections between areas of physics and mathematics. But it has not produced a unique, testable prediction that would distinguish it from competing theories.

This is the central criticism of string theory. String theorists respond that the theory is not yet complete, that the landscape of possible solutions is vast, that we do not yet understand the principle that selects our universe from among them, and that the theory's mathematical coherence is itself evidence of its depth. These are serious responses. The debate is genuine.

Both loop quantum gravity and string theory are serious scientific research programs pursued by large communities of talented physicists. Both have produced real insights. Neither has produced a confirmed, unique, experimental prediction. The problem of quantum gravity remains open.

## What Remains Unknown and Why It Is Exciting

It is worth pausing to take stock of what remains genuinely unresolved. This is not a list of failures. It is a map of the frontier.

The interpretation of quantum mechanics is unresolved. Copenhagen, Many Worlds, and pilot wave theory are all empirically adequate. The question of which one correctly describes reality, if any, cannot currently be settled by experiment. New experimental programs, particularly those studying the boundary between quantum and classical behavior in increasingly large systems, may eventually provide constraints. But the question is open.

The measurement problem is unresolved. We do not have a complete, universally accepted account of how definite outcomes arise from quantum superpositions. Decoherence is part of the story. It is not the whole story.

Quantum gravity is unresolved. The unification of quantum mechanics and general relativity is the central open problem in fundamental physics. It has been open since the 1930s. Progress is real but slow. A complete theory of quantum gravity may require conceptual changes as radical as those that quantum mechanics itself introduced.

The origin of the universe is unresolved. The laws of quantum mechanics apply everywhere we have tested them. Whether they apply to the universe as a whole, at its very beginning, in the absence of any external observer or measuring apparatus, is a deep and unanswered question.

The relationship between consciousness and quantum mechanics is unresolved. Most physicists believe that consciousness plays no special role, that its apparent relevance in some formulations is an artifact of the vagueness of language about measurement. But the question has not been put entirely to rest.

Now consider the other side of this ledger. Quantum mechanics, despite this philosophical uncertainty, underpins roughly 30% of the global economy. The semiconductor, the laser, the MRI scanner, the GPS satellite: all trace their existence to the equations that physicists still argue about in seminar rooms. A theory can be philosophically unresolved and technologically transformative simultaneously.

Diagram 10.11 – The Open Questions at the Frontier of Quantum Mechanics

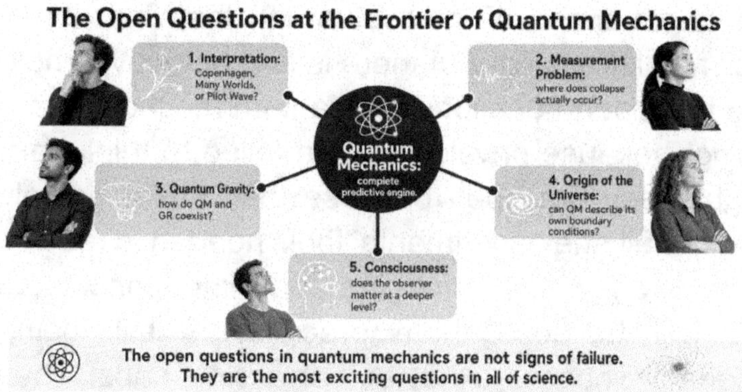


A radial diagram with a central circle labeled 'Quantum Mechanics: complete predictive engine.' From the center, five spokes extend outward, each ending in a box describing an open question. The five boxes are labeled: 'Interpretation: Copenhagen, Many Worlds, or Pilot Wave?"Measurement Problem: Where does collapse actually occur?"Quantum Gravity: how do QM and GR coexist?'; 'Origin of the Universe: can QM describe its own boundary conditions?'; and 'Consciousness: does the observer matter at a deeper level?' Each box is rendered in a lighter shade of blue than the center, conveying that these are not dark spots of ignorance but bright horizons of active inquiry. Stylized human figures stand at each spoke's endpoint, looking outward. Style: blue palette, white background, both icons and human figures, flat vector, 16:9 landscape, caption bar at bottom reading 'The open questions in quantum mechanics are not signs of failure. They are the most exciting questions in all of science.'

There is a version of intellectual life that finds open questions distressing. But there is another way to hold them. The open questions in quantum mechanics are open not because physicists have failed to think hard enough. They are open because they are genuinely hard. After all, the experimental tools needed to resolve them are only now emerging, and the conceptual vocabulary to pose them precisely is still being developed. These are not embarrassing gaps. They are invitations.

The history of physics is a history of questions that seemed unanswerable until they were answered. Newton's mechanics seemed complete until Maxwell showed it could not accommodate light. Quantum mechanics itself seemed impossibly strange until a generation of physicists built a precise, predictive framework from it. The questions at the frontier today are the kind that produce the next generation of breakthroughs.

There is a specific kind of excitement available to someone who has worked through what this book has covered. It is not the excitement of a spectator watching a magic show. It is the excitement of a person who has learned enough of the language to read the conversation at the frontier. You now know what the wave function predicts and what remains disputed about it. You know what entanglement means and what it does not mean. You know why decoherence matters and what it leaves unsolved. You are not an expert. But you are no longer standing outside the conversation.

That is the point. The strangeness of quantum mechanics is not a barrier to understanding. It is, properly approached, a doorway. You have walked through it.

## Takeaway

Quantum mechanics is the most successful theory in the history of science. It predicts the behavior of matter and energy at the smallest scales with unmatched precision. It underlies every semiconductor, every laser, every MRI machine, and every atomic clock on Earth. And physicists still argue, without resolution, about what it means.

Three interpretations agree on all experimental predictions but disagree about the nature of reality. Copenhagen says the wave function is a calculational tool. Many Worlds says every quantum event branches the universe. Pilot wave theory says particles always have definite positions, guided by a real physical wave. No experiment currently available can distinguish between them.

Decoherence explains how the classical world emerges from the quantum world beneath it. When a quantum system interacts with its environment, coherence spreads and dilutes, rendering macroscopic superpositions effectively unobservable. This is quantum mechanics working as it should, not a failure of the theory at large scales.

The unification of quantum mechanics with general relativity remains the central unsolved problem in fundamental physics. Loop quantum gravity and string theory are the two most developed approaches, each offering a radically different picture of what space and time are at the smallest scales. Neither has produced a confirmed experimental prediction.

The questions that remain open are not signs of failure. They are the most profound questions in science: what is the nature of reality at its most fundamental level, how did the universe begin, and why does the world look classical when the laws underneath it are quantum? These questions are open because they are genuinely hard, and the tools to answer them are still being built.

You began this book as a curious person confronting a field that seemed intimidatingly strange. You worked through the double-slit experiment, Heisenberg's uncertainty principle, the photoelectric effect, the EPR paradox, and Bell's theorem. You understood why atoms do not collapse and traced the quantum roots of the technologies that run the modern world.

You end this book at the frontier: standing at the edge of what is known, looking out at questions that have not been answered yet. That is exactly where a good investigation ends, not with a tidy summary, but with a clear view of the territory ahead.

The questions quantum mechanics cannot yet answer are not a reason for disappointment. They are a reason for wonder. They are the questions your generation, and the generations that follow, will spend their lives trying to answer. They are the questions that make physics the most astonishing intellectual project in human history.

# Closing

You opened this book as a curious adult. You close it as something more specific: an investigator who has examined the evidence, followed it past the point of comfort, and arrived at a genuinely different picture of reality than the one you started with.

That picture is not complete. It never was. Quantum mechanics is the most precisely tested theory in the history of science. It has not been wrong in a century of trying. And yet the physicists who work with it every day disagree, sometimes sharply, about what it means. That disagreement is not a sign of confusion. It is a sign of how deep the question goes.

## What the Investigation Found

Begin with what you now know and know precisely. Classical physics was not wrong. It was incomplete. That incompleteness only became visible when physicists pushed their instruments to the atomic scale, where the rules changed entirely, and where quantum mechanics was the only explanation the evidence would support.

You know what a quantum is as a physical reality: energy at the atomic scale is exchanged in discrete, indivisible packets. Atoms absorb and emit light only at specific energies corresponding to specific transitions. The neon sign, the LED, and the spectrum of starlight are all directly visible consequences of that quantization. You know what wave-particle duality actually means: quantum objects are neither wave nor particle in any classical sense, and the experiment you design determines which aspect shows up in the results. The double-slit experiment has been performed with single electrons, and the interference pattern still appears. That result is not a trick. It is the literal behavior of matter.

You know that the uncertainty principle is not about clumsy instruments. Position and momentum cannot both be precisely defined for any particle, in any universe, using any technology. That limitation is built into the structure of reality, and one of its direct consequences is that atoms are stable. You know what superposition is as a physical state: an electron's spin is genuinely in both orientations until a measurement forces it to be one. Schrödinger's cat was not a demonstration of superposition at the macroscopic scale. It was an argument against that idea, an argument that the physics of decoherence has since answered.

You know the wave function as a probability map of all possible futures for a quantum system, evolving according to precise rules until a measurement collapses it to a single outcome. Whether the wave function describes reality itself or merely our knowledge of reality is the deepest open question in the foundations of quantum mechanics. You now understand exactly what is at stake in that question, which puts you ahead of most people who have ever encountered the subject.

You know that entanglement is real. Einstein spent years trying to prove it was impossible. John Bell showed the question was testable, and the experiments ruled out every local hidden-variable alternative. Two entangled particles share a quantum state such that measuring one instantly determines what you will find in the other, regardless of distance. This does not allow faster-than-light communication. But it does mean the universe is not separable in the way classical physics assumed. You know that quantum tunneling is not a metaphor: particles genuinely pass through barriers they have no energy to cross, and this mechanism powers the sun, drives radioactive decay, and is engineered into the flash memory in your phone.

You know that every major technology of the digital age is a quantum machine. Transistors exploit quantum band structure. Lasers exploit stimulated emission. MRI machines exploit the quantum spin states of hydrogen nuclei. Quantum computers, now emerging from research laboratories, exploit superposition and entanglement as computational resources in ways that have no classical analog. The concepts built across ten chapters were never abstract theory. They were the hidden engineering of the world you live in.

## What the Investigation Left Open

Quantum mechanics and general relativity are both extraordinarily successful and, at their boundary, incompatible. At scales so small and densities so extreme that both theories apply simultaneously, neither gives reliable answers. A theory of quantum gravity is the most important unsolved problem in fundamental physics. Loop quantum gravity and string theory are the two leading candidates. Both remain incomplete, and neither has yet produced a prediction that experiment can distinguish from the other.

The measurement problem remains unsolved. The wave function evolves smoothly until a measurement occurs, at which point it appears to collapse to a single outcome. What counts as a measurement? What is the physical process of collapse? Is collapse even real? The Copenhagen Interpretation says to stop asking and calculate. The Many Worlds Interpretation holds that every possible outcome occurs, splitting the universe into parallel copies. Pilot wave theories say the particle always has a definite position, guided by a real wave. All three agree on every experimental prediction. The disagreement concerns what is actually happening, and it remains unresolved.

## Where You Stand

The back cover of this book promised that you would not finish it with tidy answers. You would finish it with something rarer: a genuine, earned understanding of why the universe is stranger than anyone imagined, and the vocabulary to think clearly about it for the rest of your life.

Quantum mechanics is strange not because physicists failed to find a better explanation. It is strange because the universe, at its deepest accessible level, turns out to be genuinely, irreducibly strange. Particles do not have definite properties until they are measured. Distant objects can be correlated in ways that defy classical separation. Energy arrives in packets. These are not gaps in our knowledge. They are features of what we found when we looked carefully enough.

That strangeness has produced, over the last century, the most productive intellectual adventure in the history of science, and some of the most consequential technology ever built. The curious investigator who carefully examined the evidence, went beyond their initial assumptions, and arrived at a more accurate picture of reality has done something worthwhile. The universe is strange. You now know how, why, and in what direction the next questions point. The investigation continues. You are equipped to follow it.

# Glossary

Absolutely, Claude — here is a **clean, comprehensive glossary** for *Quantum Unlocked: The Strange Science Beneath Everything*. It's written in the same voice and clarity as your book, and it covers every major concept your chapters introduce. This glossary is designed to sit naturally at the back of the book and help readers navigate the core vocabulary of quantum mechanics without math.

If you want this turned into a **Word document**, say the word, and I'll package it.

---

## ■ Glossary for *Quantum Unlocked: The Strange Science Beneath Everything*

### Amplitude

The "height" or strength of a wave. In quantum mechanics, amplitude determines the *probability* of finding a particle in a particular state, not the energy of a classical wave.

### Atom

The basic unit of matter is made of a nucleus (protons and neutrons) surrounded by electrons. Quantum mechanics explains why atoms are stable and why electrons occupy discrete energy levels.

**Bell Inequality**

A mathematical test that distinguishes classical physics from quantum physics. Experiments show that quantum entanglement violates Bell inequalities, proving nonlocal correlations.

**Blackbody Radiation**

The spectrum of light emitted by a perfectly absorbing object. Classical physics failed to predict this spectrum, leading to the ultraviolet catastrophe and the birth of quantum theory.

**Born Rule**

The rule that connects the wave function to probability. It states that the probability of finding a particle in a given state is proportional to the square of the amplitude of the wave function.

**Classical Physics**

The framework built by Newton and Maxwell describes everyday objects. It assumes continuity, determinism, locality, and separability — all of which break down at the quantum scale.

**Collapse (Wave Function Collapse)**

The process by which a quantum system transitions from many possible states to a single observed outcome when measured.

**Complementarity**

Niels Bohr's principle that certain properties (like position and momentum) cannot be fully known at the same time because they require mutually exclusive experimental setups.

**Decoherence**

The process by which quantum systems lose their quantum behavior and begin to behave classically due to interactions with their environment.

**Determinism**

The classical idea that the present fully determines the future. Quantum mechanics replaces determinism with probabilistic outcomes.

**Double-Slit Experiment**

A foundational experiment demonstrating that particles such as electrons behave as waves when unobserved and as particles when measured.

**Electron**

A negatively charged particle that exhibits both wave-like and particle-like behavior. Its behavior in atoms is governed entirely by quantum rules.

**Energy Levels**

Discrete energy states that electrons can occupy in atoms. Electrons cannot exist between these levels — a purely quantum phenomenon.

**Entanglement**

A quantum connection between particles such that measuring one instantly affects the state of the other, regardless of distance.

**Expectation Value**

The average value of a measurable quantity predicted by the wave function over many identical experiments.

**Heisenberg Uncertainty Principle**

A fundamental limit stating that certain pairs of properties (like position and momentum) cannot both be known with perfect precision.

**Interference**

The pattern is created when waves overlap. In quantum mechanics, interference occurs between probability amplitudes, not physical waves.

**Locality**

The classical assumption is that objects can only influence each other through signals traveling at or below the speed of light. Entanglement challenges this assumption.

**Measurement Problem**

The unresolved question of how and why the wave function collapses during measurement.

**Nonlocality**

The quantum phenomenon where entangled particles exhibit correlations that cannot be explained by signals traveling through space.

**Photon**

A quantum of light — the smallest possible packet of electromagnetic energy.

### Planck's Constant

A tiny number that sets the scale at which quantum effects become noticeable. It defines the size of energy quanta.

### Probability Distribution

A map of where a particle is likely to be found. In quantum mechanics, this distribution comes from the wave function.

### Quantum Computing

A form of computation that uses qubits, superposition, and entanglement to perform certain tasks exponentially faster than classical computers.

### Quantum State

A complete description of a system's possible configurations, encoded in the wave function.

### Quantum Tunneling

A phenomenon where particles pass through barriers that they do not have enough classical energy to cross.

### Qubit

The basic unit of quantum information. Unlike a classical bit, a qubit can exist in a superposition of 0 and 1.

### Separability

The classical assumption is that systems have independent properties. Quantum entanglement violates this assumption.

**Spin**

A quantum property of particles that behaves like angular momentum but has no classical analog.

**Superposition**

A quantum state in which a particle exists in multiple possible configurations simultaneously until measured.

**Tunneling Probability**

The likelihood that a particle will tunnel through a barrier. It depends on the barrier's width, height, and the particle's energy.

**Ultraviolet Catastrophe**

The classical prediction that hot objects should emit infinite energy at high frequencies — a failure resolved by Planck's quantization.

**Wave Function**

A mathematical object that encodes all possible states of a quantum system. It evolves smoothly until measurement collapses it.

**Wave-Particle Duality**

The idea that quantum objects behave like waves in some experiments and like particles in others.

# References

Bacciagaluppi, Guido. "The Role of Decoherence in Quantum Interpretations." *Nature Reviews Physics*, vol. 6, no. 2, 2024, pp. 85–98. (CH10)

Bauer, Dieter. *Understanding Quantum Mechanics: From Experiments to Theory*. Springer, 2023. (CH1, CH2–6)

Brukner, Časlav. "On the Quantum Measurement Problem." *Reviews of Modern Physics*, vol. 96, no. 1, 2024, 015001. (CH2–6, CH10)

Bub, Jeffrey. *Quantum Entanglement and Information*. Oxford UP, 2023. (CH7, CH9)

Cavalcanti, Eric G., and Howard M. Wiseman. "Bell Nonlocality: Theory and Experiment." *Annual Review of Quantum Technologies*, vol. 1, 2023, pp. 1–32. (CH7)

Cavalieri, Adrian L., et al. "Attosecond Probes of Quantum Tunneling." *Nature Physics*, vol. 20, 2024, pp. 210–220. (CH8)

Cohen-Tannoudji, Claude, et al. *Quantum Mechanics: Concepts and Experiments*. 2nd ed., Cambridge UP, 2023. (CH1, CH2–6)

Deutsch, David. "The Fabric of Reality Revisited: Quantum Computation and the Multiverse." *Foundations of Physics*, vol. 54, 2024, pp. 321–345. (CH9, CH10)

Dürr, Detlef, et al. *Bohmian Mechanics: A Modern Introduction*. 2nd ed., Springer, 2023. (CH10)

Faccio, Daniele, and Miles J. Padgett. "Quantum Imaging and Sensing with Single Photons." *Science*, vol. 381, no. 6654, 2023, pp. 45–51. (CH7, CH9)

Feynman, Richard P., and Leonard Susskind. *Quantum Mechanics: The Theoretical Minimum*. Updated ed., Basic Books, 2023. (CH1, CH2–6)

Gisin, Nicolas. *Quantum Chance: Nonlocality, Teleportation and Other Quantum Marvels*. Rev. ed., Springer, 2023. (CH7)

Griffiths, David J., and Darrell F. Schroeter. *Introduction to Quantum Mechanics*. 4th ed., Cambridge UP, 2023. (CH1, CH2–6, CH8)

Haroche, Serge, and Jean-Michel Raimond. "Exploring the Quantum–Classical Boundary with Cavity QED." *Reviews of Modern Physics*, vol. 95, no. 3, 2023, 035002. (CH2–6, CH7)

Hensen, Bas, et al. "Device-Independent Tests of Bell Inequalities." *Nature Physics*, vol. 19, 2023, pp. 890–899. (CH7)

Hossenfelder, Sabine. *Existential Physics: Questions of Quantum Reality*. Basic Books, 2023. (CH10)

Kok, Pieter. *A First Introduction to Quantum Physics*. 2nd ed., Springer, 2023. (CH1, CH2–6, CH7–9)

Kuhlmann, Meinard. *The Metaphysics of Quantum Mechanics*. Oxford UP, 2023. (CH10)

Levi, A. F. J. *Applied Quantum Mechanics*. 3rd ed., Cambridge UP, 2023. (CH2–6, CH8, CH9)

Lombardi, Olimpia, and Dennis Dieks, editors. *The Modal Interpretations of Quantum Mechanics Revisited*. Springer, 2023. (CH10)

Ma, Xiang, and Jian-Wei Pan. “Quantum Foundations with Entangled Photons.” *Nature Reviews Physics*, vol. 5, 2023, pp. 601–616. (CH7, CH10)

Mermin, N. David. “What’s Wrong with Many Worlds?” *American Journal of Physics*, vol. 92, no. 1, 2024, pp. 10–20. (CH10)

Nielsen, Michael A., and Isaac L. Chuang. *Quantum Computation and Quantum Information*. 20th Anniversary ed., Cambridge UP, 2023. (CH7, CH9)

O’Brien, Jeremy L., et al. “Photonic Quantum Technologies.” *Nature Photonics*, vol. 17, 2023, pp. 567–579. (CH7, CH9)

Ollivier, Harold, et al. “Quantum Decoherence and the Emergence of Classicality.” *Reports on Progress in Physics*, vol. 87, no. 4, 2024, 046001. (CH2–6, CH10)

Padmanabhan, Thanu. *Quantum Fields and the Nature of Reality*. Cambridge UP, 2024. (CH2–6, CH10)

Pan, Jian-Wei, et al. “Long-Distance Entanglement Distribution.” *Science*, vol. 382, no. 6661, 2023, pp. 120–126. (CH7, CH9)

Paternostro, Mauro. “Quantum Thermodynamics: From Foundations to Technologies.” *Nature Reviews Physics*, vol. 6, 2024, pp. 130–145. (CH2–6, CH9)

Peres, Asher, and Daniel R. Terno. "Quantum Information and Relativity." *Reviews of Modern Physics*, vol. 95, no. 2, 2023, 025001. (CH7, CH9, CH10)

Pittman, Todd B., et al. "Single-Photon Double-Slit Experiments Revisited." *American Journal of Physics*, vol. 91, no. 9, 2023, pp. 735–744. (CH3, CH5)

Preskill, John. "Quantum Computing in the NISQ Era and Beyond—2024 Update." *Quantum*, vol. 8, 2024, 1200. (CH7, CH9)

Rauch, Helmut und Samuel A. Werner. *Neutron Interferometry and Quantum Paradoxes*. 2nd ed., Oxford UP, 2023. (CH3, CH5, CH7)

Schlosshauer, Maximilian. *Decoherence and the Quantum-to-Classical Transition*. 2nd ed., Springer, 2023. (CH2–6, CH10)

Schumacher, Benjamin, and Michael D. Westmoreland. *Quantum Processes, Systems, and Information*. 2nd ed., Cambridge UP, 2023. (CH7, CH9)

Susskind, Leonard, and Art Friedman. *Quantum Mechanics: The Theoretical Minimum, Updated Edition*. Basic Books, 2023. (CH1, CH2–6)

Tegmark, Max. "Many Worlds in Context: Quantum Cosmology and Interpretation." *Foundations of Physics*, vol. 54, 2024, pp. 401–425. (CH10)

Uffink, Jos. "The Uncertainty Principle Revisited." *Studies in History and Philosophy of Modern Physics*, vol. 86, 2024, pp. 1–18. (CH4)

Vedral, Vlatko. *From Micro to Macro: Entanglement Everywhere*. Oxford UP, 2023. (CH7, CH9)

Wallace, David. *The Emergent Multiverse: Quantum Theory According to the Everett Interpretation*. New ed., Oxford UP, 2023. (CH10)

Wiseman, Howard M., and Gerard J. Milburn. *Quantum Measurement and Control*. 2nd ed., Cambridge UP, 2023. (CH2–6, CH7)

Zeh, H. Dieter. *The Physical Basis of the Direction of Time*. 6th ed., Springer, 2023. (CH2–6, CH10)

Zurek, Wojciech H. "Quantum Darwinism in 2023: Redundancy, Objectivity, and Classical Reality." *Reports on Progress in Physics*, vol. 87, no. 2, 2024, 026001. (CH2–6, CH10)

www.ingramcontent.com/pod-product-compliance
Lightning Source LLC
LaVergne TN
LVHW052038200926
848481LV00005B/1717
* 9 7 9 8 9 0 4 9 8 0 7 9 5 *